How to Pass the APC: Essential Advice for General Practice Surveyors

Austen Imber

2004

Routledge
Taylor & Francis Group

LONDON AND NEW YORK

ISBN 0 7282 0429 0

Typeset by Amy Boyle, Rochester, Kent

Contents

Foreword

The RICS Assessment of Professional Competence is commonly cited by graduates as an area of study in which a practical insight is required in the final assessment process.

This view has informed our regular coverage of the APC in the Mainly for Students section of the *Estates Gazette* over the past decade. This section of the magazine has always scored highly with readers who clearly value the clear practical advice that is given. Therefore we are now pleased to publish *How to Pass the APC*.

This complements the material available from RICS, and shows APC candidates how, with the right commitment, they can succeed at their first attempt. This is not only through considering the many facets of final assessment, but taking the right approach to many other aspects throughout their professional training period.

Peter Bill
Editor *Estates Gazette*
February 2004

Preface

How to Pass the APC: Essential Advice for General Practice Surveyors shows general practice surveyors how they can most effectively apply themselves to the APC – and be proud to secure MRICS status at their first attempt. This includes helping candidates understand what they need to know in their subject areas to be successful with the RICS APC interview.

The national pass rate for general practice surveyors taking the final assessment interview has been around 65% over recent years (which after allowing for re-sits, equates to an approximate 50% first time pass rate). This is low, considering that one of the elements of the APC is that employers (as supervisor and counsellor) are required to declare that candidates have reached the necessary levels before putting them forward for final assessment.

Reasons for the number of unsuccessful candidates include a lack of awareness as to what exactly needs to be understood within competency areas, supervisors and counsellors signing off candidates despite not having met the required levels, and candidates being prepared to take a speculative gamble on APC success – some consoling themselves that failure would still be good experience for further attempts. A related factor is candidates often focusing on two months of revision, rather than at least two years of concerted learning.

It is also important for candidates to demonstrate capability to assessors in following RICS APC requirements – such as competency selection, professional development/CPD activities, and the structure and finer content of written submissions, in particular the critical analysis.

Even where candidates meet the required standards in complying with RICS requirements, and gaining the knowledge and understanding of their subject areas, they can struggle to perform well under pressurised interview conditions.

How to Pass the APC draws on the experiences of APC candidates, RICS APC assessors and employers providing work experience and related structured APC training programmes. It complements the guidance made available to candidates by RICS, and repeats only essential technical information.

Estates Gazette's 'Mainly for Students' series provides ongoing coverage of aspects of the APC. These and the more recently commenced 'Mainly for Students education and training updates', prepared in conjunction with RICS, have seen GVA Grimley, Advantage West Midlands and Birmingham Property Services featured on a number of occasions. As well as the support provided for their own candidates, these employers have facilitated courses and other support for APC candidates. The individuals behind their firms' success, and who have contributed to *How to Pass the APC*, are:

Scott Kind, UK training and development manager, GVA Grimley, international property consultants. Scott formerly worked with Nat West and Financial Times, and is responsible for the implementation of GVA Grimley's APC training programme, as well CPD and business skills training.

Kitt Walker, Advantage West Midlands regional development agency, supported also by Honor Boyd, Karen Yeomans and Yaseen Mohammed (since moved to Yorkshire Forward). As part of the Advantage West Midlands CPD Centre initiative, the agency has facilitated a range of training opportunities for surveyors, some of which promote the opportunities in regeneration areas to developers, investors and occupiers.

Jacky Gutteridge, training and development manager, Birmingham Property Services, the in-house property consultancy of Birmingham City Council. Jacky is responsible for the implementation of BPS's APC programme, which within its first phase between January 1999 and April 2003, secured a 13 out of 13 first time pass rate.

The work of Margaret Harris, RICS CPD Officer, in respect of lifelong learning is also acknowledged. This includes the extracts from RICS material in the section 'Beyond the APC'.

Other contributions are made by James Turner, an APC candidate with Saxon Law; Paul Richardson, a surveyor and former APC candidate with GVA Grimley; Claire Paraskeva, an associate and APC assessor with GVA Grimley; and Mark Clarke, a surveyor with Advantage West Midlands.

Also acknowledged is the support of Peter Bill, *Estates Gazette* editor, and Alison Richards, *Estates Gazette Books* Commissioning Editor, plus all those involved in the process – including Rebecca Chakraborty of EG Books, and Adam Tinworth and Phil Brown regarding the coverage of the APC in *Estates Gazette*'s 'Mainly for Students' series. Essential support in proof reading has been provided by Howard Imber, and in production by Audrey Andersson and Amy Boyle.

Thanks are due to Midlands Property Training Centre for their co-ordination of material, and ongoing help to general practice graduates as an independent facilitator of training support.

Austen Imber
April 2004

Chapter 1

APC Overview

The Assessment of Professional Competence (APC) commenced in 1992. It typically involves candidates gaining a professional training period of at least two years, and undertaking a final assessment process to determine whether they have reached the necessary standard to become chartered surveyors (and benefit from the designation MRICS – Member of the Royal Institution of Chartered Surveyors).

Most candidates commence the APC after completing their property degree at one of the RICS partnership universities. Non-cognate degree holders can now take the APC concurrently with an RICS approved part-time property degree, and sandwich students may also commence the APC in their placement year. There are also special routes to chartered surveyor status, such as where 10 years or more of relevant work experience has been gained.

RICS organises the APC around the way the property industry is structured, and the type of work in which graduate surveyors are often involved. Ongoing refinements have been made over the years, such as recent changes to reflect the increasing number of candidates specialising in the telecoms sector.

APC route

Candidates' area of practice is designated by an 'APC route'. General practice surveyors typically select the Commercial Real Estate Practice/Commercial Property route (the term 'Commercial Property' being used in *How to Pass the APC*). For candidates involved mainly in property development work, the Planning and Development route may be taken. Although a Valuation route is available, few candidates take this.

Competencies

Over their training period, candidates develop skills in areas of practice known as 'competencies'. These comprise mandatory competencies, core competencies and optional competencies. Mandatory competencies represent the skills required by all surveyors, and are the same for all APC routes. Core competencies relate to candidates' APC route, and optional competencies are selected by candidates having regard to the precise areas of work in which they are involved.

Chapter 2, Starting the APC, outlines the requirements in more detail, and reference should be also made to the relevant RICS APC guides.

Levels of attainment

Levels of attainment need to be reached by candidates in their competencies. The RICS APC guides state that level 1 is 'knowledge and understanding', level 2 is 'application of knowledge and understanding', and level 3 is 'reasoned advice and/or depth of technical knowledge'.

Diary and log book

Day-to-day experience is recorded in a 'diary'. Each half-day or full-day is allocated to a core or optional competency, and totalled in a 'log look'. The log book, therefore, shows the experience being gained on a month-by-month basis, with periodic grand totals highlighting the need for wider experience (or sometimes the need to allocate work experience to competencies more accurately).

Supervisor and counsellor

Each candidate has an 'APC supervisor' and an 'APC counsellor'. The supervisor is usually an immediate manager who oversees day-to-day work, and the counsellor is a manager able to provide a broader overview of candidates' process, and, for example, facilitate opportunities for wider experience to be gained. Precise arrangements and responsibilities vary depending on the nature of the employer, and more detail is provided in Chapter 3, Key Input from Supervisor and Counsellor.

A three-monthly meeting is held with the supervisor, and a six-monthly meeting with the counsellor, and progress reports are subsequently completed. The 'record of progress' involves the supervisor and counsellor signing off their candidate to the levels of attainment considered to have been achieved in their competencies.

Professional development

In addition to their case work, candidates undertake surrounding learning known as 'professional development'. This is the equivalent of 'continuing professional development' ('CPD') undertaken by qualified surveyors and comprises courses, seminars, private study, etc. A minimum of 48 hours per year is undertaken, and is detailed in candidates' 'professional development record'. Further detail is outlined in Chapter 2.

Interim assessment

After 12 months' experience, candidates undertake interim assessment. This involves completing the diary, log book and professional development record, and also preparing a 'summary of progress', and 'forward plan'.

The summary of progress is a 2,000 word outline of the work experience gained over the first year, and the forward plan is a 1,000 word overview of the experience and skills to be developed over the second year.

Interim assessment is completed internally with the supervisor and counsellor, and although not submitted to RICS at the interim stage, is included as part of the final assessment submissions.

Final assessment

Final assessment comprises written submissions, and a final assessment interview.

Submissions include a 'critical analysis', a 2,000 word summary of progress for the period since interim assessment, the log book, professional development record, and the copy of the interim assessment (as well application forms, etc). Candidates who have been referred previously will have slightly different requirements. Special comments for referred candidates are included in Chapter 12, Miscellaneous. ('Refer' and 'referred' is RICS terminology for 'fail' and 'failed').

The critical analysis is a 3,000 word report on a case in which candidates have been actively involved. This is an important demonstration of candidates' ability to perform in practice, and forms a 10 minute presentation as part of the final assessment interview. It is also the subject of around 10–15 minutes of interview questioning. The summary of progress is similar to the report undertaken at the interim stage, and outlines the further experience that has been gained. Interim assessment and final assessment also involve the completion of the supervisor's and counsellor's reports, and bringing the record of progress up to date.

Neither the diary nor the three-monthly and six-monthly reports are submitted to RICS as part of final assessment, although the supervisor's and counsellor's reports which accompany interim and final assessment summaries of progress will be submitted. Although certain elements are not submitted to RICS, they may be called up by assessors at final assessment, or may be requested by RICS as part of spot-checks at any time.

The final assessment interview is undertaken by candidates with three RICS 'assessors', and broadly comprises a five-minute introduction, the candidate's 10-minute presentation on their critical analysis and 45 minutes of interview questioning. Assessors will then consider whether the candidate has passed or should be referred, and the candidate will be informed of the decision by RICS 21 days later. For general practice surveyors, the interviews are held in March/April and September/October each year at various venues around the country.

The final assessment stage of the APC is geared towards the work experience gained by candidates. Although the APC may appear complex on an initial examination of the RICS guides, this is necessarily so in order that every candidate can be assessed with regard to the nature of their employer's business, and the day-to-day case work in which they are involved. There are still, however, areas of knowledge beyond day-to-day work that are required in order to become a chartered surveyor.

All candidates have gaps in experience to some degree, and some seek to redress this more pro-actively than others. The importance of candidates viewing the APC as a training period of two years (rather than a revision period of two months) cannot be stressed too strongly. It is also, of course, important that candidates select an employer who is able to provide appropriate work experience.

Structured training agreement

Employers need to have a structured training agreement in place (approved by an RICS Regional Training Adviser) before their candidates can formally begin their professional APC training period. This provides general information about the employer and the nature of their business, and specific information on the experience which is available, including in relation to APC routes and competencies.

The content of the documentation varies between large practices with many recruits per year, and smaller concerns taking on a graduate only infrequently. As well as the documentation having to meet RICS requirements, some firms use this as a tool to set out their internal APC processes to all their candidates, supervisors and counsellors.

Different APC systems

The APC is refined on an ongoing basis to keep up-to-date with market developments, and also any changes in RICS corporate structure. Sometimes the changes are minor, whereas on other occasions, more extensive changes mean that different candidates can be working to different systems.

How to Pass the APC: Essential Advice for General Practice Surveyors works to the APC system effective from August 2002. This is covered by the RICS guides dated July 2002 as edition 1 (noting that further editions could be introduced to deal with variations). There may also be candidates using *How to Pass the APC* who commenced the APC prior to August 2002, and may be using the old 5th edition, or earlier, guides. The old system adopted common competencies instead of mandatory competencies, and although many changes took place throughout the system, the requirements affecting general practice surveyors did not change too substantially. Key differences include the availability of a wider range of optional competencies than previously prescribed for individual APC routes, and the renaming of some competencies (such as 'Property Marketing' to 'Purchase, Disposal and Leasing', and 'Estate Management' to 'Real Estate Management').

Candidate responsibility

It is important that candidates establish the precise requirements which affect them personally, and do not rely on the comments of other candidates, or managers, who may be familiar only with other systems. Candidates who have been referred will be subject to particular requirements.

Special APC routes, such as experienced, expert and academic/research, all have their particular eligibility criteria, and other requirements.

Other RICS personnel involved

In addition to supervisors, counsellors and assessors, candidates may also deal with the RICS APC staff, RICS APC Doctors and RICS Regional Training Advisers.

The RICS APC staff are based in Coventry, and administer all stages of the APC. They are available to deal with queries, although it is important that candidates examine the RICS guides in detail themselves, rather than telephoning RICS for a personal guide through the system.

The RICS APC Doctors are local chartered surveyors available to assist APC candidates. The RICS Regional Training Advisers help RICS implement education and training initiatives on a regional basis, including in respect of the APC.

Training support

The larger firms are likely to have training managers who run established in-house APC training programmes. This will include support from the firm's chartered surveyors who are also RICS APC assessors, with assistance also being sought from external training providers (such as through in-house seminars or consultancy services). Candidates in smaller organisations may find that there are no colleagues who are familiar with the APC, and that they must consequently take greater responsibility for ensuring that RICS APC requirements are met.

There are various seminars and day courses run by RICS, universities (such as Kingston University) and commercial training providers that help guide candidates through the APC. Most events, however, concentrate on final assessment. RICS guides and supplementary material, including publications such as *How to Pass the APC*, and *The APC – your practical guide to success* (written by John Wilkinson and published by RICS) tend to be sufficient to get candidates started.

Contributors to this book, Advantage West Midlands, the regional development agency for the West Midlands, facilitates free of charge training support for APC candidates as part of their 'Advantage West Midlands CPD Centre initiative'. Midlands Property Training Centre, an independent facilitator of training for general practice surveyors, similarly help ensure that free of charge APC and CPD is made available to APC candidates. (Current details can be received by requesting 'latest GP APC information' from apc@mptcentre.org.)

Some candidates indicate that they are not being provided with information in respect of APC support, etc. However, training providers rely largely on the RICS for providing a data base of candidates' names and work addresses. Those candidates opting not to receive details from third parties consequently deny themselves the full range of APC support available.

Chapter 2

Starting the APC

To summarise from the overview of the APC in Chapter 1, at the commencement of their APC training period, candidates will need consider the following aspects, each of which are examined individually below. Candidates should refer to the RICS APC guides which provide the supporting technical information (obtainable from RICS: Tel 020 7222 7000, and via the RICS website, www.rics.org).

- Registering for the APC.
- APC route.
- Supervisor and counsellor input.
- Competency achievement planner.
- Mandatory competencies.
- Core competencies.
- Optional competencies.
- Diary recording and log book entries.
- Structured training agreement.
- Progress reports and the record of progress.
- Professional development.

Registering for the APC

In order to begin their training period, candidates must formally register with RICS, and be given their registration date. Delays in doing so could result in final assessment having to be undertaken later than planned. Within the current RICS rules, if candidates register before the end of October, they will be eligible for final assessment in September/October two years later. This assumes a training period unbroken by a sandwich course or other reason, and that candidates meet the necessary levels, as certified by their supervisor and counsellor. There could be different arrangements applying to individual candidates, such as expert, experienced and academic/research route candidates. Also, although RICS may take several weeks to progress the application, the registration date will be back-dated to the date of receipt. On beginning work, candidates should look to register for the APC as soon as possible in order that the maximum eligible period of experience can be gained.

The rules stated in the RICS APC guides are often subject to finer requirements. The guides state, for example, that the APC will normally consist of a minimum of 24 months of training. However, candidates can commence their APC training period at the end of October and take the interview at the start of September just under two years later. Such nuances do not tend to cause difficulties for candidates, as RICS correspondence with candidates will set out more detailed individual guidance than is feasible in the RICS guides. The guides also state that candidates will need to have gained a total of 400 days' experience before being eligible to sit.

The majority of the registration requirements are straightforward, but in order to complete registration, candidates must determine their APC route and APC competencies, and complete a competency achievement planner. The RICS APC guides and RICS website include copies of the templates to which candidates will work, as well as guidance on most aspects of the APC.

APC route

As mentioned on page 1, general practice surveyors typically select the Commercial Real Estate Practice/Commercial Property route (the term 'Commercial Property' being used in *How to Pass the APC*). This would be suitable for candidates involved in general practice work, comprising capital valuations, rent reviews, lease renewals, assignments, sub-lettings, service charge management, rent collection and other tenant default, lettings, sales, purchases, rating, etc. Development work could also be undertaken, and may incorporate planning, development appraisal, economic development/regeneration and compulsory purchase work.

General practice surveyors involved in planning and development work may take the Planning and Development route. This tends to be taken by candidates heavily involved in development work, especially planning, and not also involved in the broader general practice areas mentioned above. The Commercial Property route is sometimes still considered suitable for candidates involved in development work. One reason is that at final assessment, candidates select a 'specialist area', which could be 'Development and planning'. This enables RICS to ensure that candidates' areas of experience are represented on the interview panel, and generally helps steer certain lines of questioning, in order to reflect candidates' experience. Competency requirements for the respective routes, as indicated on page 12, can also influence the selection of the appropriate APC route.

The Valuation route can be taken by candidates who are involved predominantly in valuation work, but few candidates select this. The Management Consultancy route may be suitable for the relatively few candidates involved in corporate property consultancy roles. Residential Real Estate Practice or Residential Survey may occasionally be taken by candidates with a general practice background, but who have become more involved in residential work.

The RICS guides provide a commentary on the type of work in which candidates will be involved under each APC route. This also provides a helpful overview of the work undertaken by other types of surveyors.

Supervisor and counsellor input

On receiving the relevant information pack from RICS, including APC guides, application forms, etc., candidates should work closely with supervisors and counsellors on understanding how the APC works, and how its various requirements fit in with the candidate's circumstances.

For supervisors and counsellors having already gained experience of the APC, either as managers or former APC candidates, the task is relatively straightforward.

However, supervisors and counsellors need to be familiar with the APC system to which candidates are working, noting the periodic changes to the system. Supervisors and counsellors could be responsible for a number of candidates working to different systems and having different requirements.

For managers without any experience of the APC, time needs to be taken to familiarise themselves with the APC requirements applying to the candidate. However, it is not the responsibility of the supervisor, or counsellor, to learn about the APC in detail, to then instruct their candidate what to do. Responsibility for establishing the APC requirements rests with candidates, with managers' support concentrated on aspects with which candidates tend to struggle on commencing employment. One key area is for candidates to find out from their supervisor the type of work which may generally fall into each competency, and how this relates to the type of experience that the employer is able to provide.

The supervisor will usually be the candidate's immediate manager, and the counsellor someone more senior who is able to take a wider overview of the candidate's training, including any opportunities to gain wider experience. In a smaller firm, this could be by providing varied case work, or in a larger firm could involve a change of department.

Further detail on the key input of supervisors and counsellors is set out in the next chapter.

Competency achievement planner

The competency achievement planner is an aspect of the initial registration process, but serves also as an ongoing facility to monitor the experience being gained, and the attainment levels reached by candidates.

Under each of the mandatory, core and optional competencies (see below), candidates will chart their anticipated progress over their two-year training period. Although in many cases, case work is not known at this stage, and it may not be possible to even determine some competencies, completion of the competency achievement planner with the supervisor and counsellor is still a useful process to help all parties focus on the RICS requirements. APC registration will not be complete without the competency achievement planner.

The competency achievement planner completed as part of initial registration will not be made available by RICS to the assessors as part of final assessment, and changes to competencies and levels of attainment can generally be made throughout the training period.

As indicated in Chapter 1, levels 1, 2 and 3 (for all competencies) are defined by RICS as follows:

Level 1 – Knowledge and understanding.
Level 2 – Application of knowledge and understanding.
Level 3 – Reasoned advice and/or depth of technical knowledge.

Experience does not necessarily need to be gained to reach level 1, but will be necessary to reach level 2. Level 3 is achieved through a detailed working knowledge, including the provision of advice to clients/employers. Level 2 can be sometimes be achieved in a more indirect way than level 3. One example would be an investment agent or valuer taking the optional competency of Landlord and Tenant to level 2 by considering the detailed effect of lease terms on capital/investment value. Candidates would still need to be aware of rent review and lease renewal processes, and although there may be relatively little, if any, sole responsibility for rent review or lease renewal cases, the necessary level of attainment could be gained by shadowing colleagues in cases, and through professional development activities (such as seminars and private study).

Mandatory competencies

Mandatory competencies represent the general skills required by surveyors. These were introduced under the APC system effective from August 2002, and are the equivalent of 'common' competencies, which some candidates will still be working to.

All candidates, throughout all APC routes on the post-August 2002 system, are required to develop skills in the following mandatory competencies, to the stated levels:

* Code of Conduct, Professional Practice and Bye-Laws (level 3).
* Conflict Avoidance, Management and Dispute Resolution Procedures (level 1).
* Collection, Retrieval and Analysis of Information and Data (level 1).
* Customer Care (level 2).
* Environmental Awareness (level 1).
* Health & Safety (level 1).
* Information Technology (level 1).
* Law (level 1).
* Oral Communication (level 2).
* Self Management (level 3).
* Teamworking (level 1).
* Written/Graphic Communication (level 2).

All candidates must additionally achieve two further business, management, financial and interpersonal related competencies. Details are included in the RICS guides, and comprise competencies such as Negotiating Skills, Business Management, Marketing, and Corporate and Public Communications. There are also further requirements placed on candidates in certain circumstances, such as having had 10 years' experience as part of the experienced route. Again, detailed reference to the RICS guides is important.

The copy of an APC candidate's summary of progress at interim assessment is included in Chapter 4, and shows some of the areas of practice which relate to the mandatory competencies.

Core competencies

The core competencies are specific to each APC route.

For the Commercial Property route, they are a minimum of:

- Valuation (level 3).
- Inspection (level 2).
- Measurement (level 2).

For the Planning and Development route, they are a minimum of:

- Development Appraisals (level 3).
- Law (level 3).
- Mapping (level 1).
- Measurement (level 2).
- Planning (level 3).
- Valuation (level 2).

Optional competencies

Optional competencies are selected by candidates in relation to the areas of work in which they are involved. The RICS APC guides include the complete list.

For both the Commercial Property route and the Planning and Development route, candidates must select a minimum of either three competencies to level 3, or two competencies to level 3 plus two competencies to level 2.

For example, optional competencies often taken by candidates in the Commercial Property route are:

- Development Appraisals.
- Landlord and Tenant.
- Local Taxation/Assessment (rating).
- Planning.
- Purchase, Disposal and Leasing.
- Real Estate Management.

Under the pre-August 2002 system, Property Marketing is the equivalent to Purchase, Disposal and Leasing, and Estate Management is the same as Real Estate Management.

Landlord and Tenant involves rent review and lease renewal, as well as a general understanding of the effects of lease terms on rental and capital value, and landlord's estate management/tenant's occupational interests. Real Estate Management involves property management areas such as alienation (assignments and sub-letting), rent default/pursuing rent arrears, other tenant default, dilapidations/repair, service charge management, insurance, buildings/facilities management, health and safety and compliance with legislation such as the Disability Discrimination Act. Further details on the areas of work falling under individual competencies, and guidance on what candidates need to know for final assessment, is shown in Chapter 10. The copy of a summary of progress at interim assessment in Chapter 4 also provides further guidance.

Optional competencies taken less commonly by candidates in the Commercial Property route include the following:

- Asset and Investment Management.
- Capital Allowances and Grants.
- Compulsory Acquisition and Compensation.
- Conflict Avoidance, Management and Dispute Resolution Procedures (also a mandatory competency, but see later).
- Corporate Real Estate Management.
- Corporate Recovery and Insolvency.
- Economic Development.
- Insurance and Risk Management.
- Real Estate Finance and Funding.
- Strategic Real Estate Consultancy.

Candidates in the Planning and Development route may also take optional competencies such as Purchase, Disposal and Leasing, Landlord and Tenant, Compulsory Acquisition and Compensation and Economic Development. Depending on their areas of work, they may select competencies such as Contaminated Land, Development/Project Briefs, Housing Strategy and Provision, Land Use and Diversification, and Inspection (which is not a core competency for the Planning and Development route). There are also competencies relating to environmental work. Candidates in the Commercial Property route who are involved in planning and development work may also select these optional competencies.

For both Commercial Property and Planning and Development routes, other competencies from the full list set out in the RICS guides may be appropriate. For some candidates, this could include competencies related more to maintenance and/or construction. It is important that candidates scrutinise the RICS APC guides to establish the competencies which best suit their circumstances.

It was mentioned above that the requirements in respect of competencies can sometimes influence whether candidates select the Commercial Property route or the Planning and Development route. For the vast majority of general practice candidates, the Commercial Property route is an obvious selection, and for candidates primarily undertaking development work, including active involvement in planning applications, the Planning and Development route would often be more appropriate. However, some candidates involved in planning and development work would still, sometimes, take the Commercial Property route, with a view to specialising in 'Development and planning' at final assessment. A comparison of the competencies shows that in addition to Valuation (level 3) as a core competency (and leaving aside the relatively straightforward competencies of Inspection and Measurement) optional competencies of Planning (level 3) and Development Appraisals (level 3), plus only one other (to level 3), would suffice. For the Planning and Development route, along with Valuation (level 2), Planning (level 3) and Development Appraisals (level 3) would be core competencies, leaving a further three optional competencies to be selected to level 3 (or two to level 3 plus 2 to level 2). If candidates involved in planning and development work feel that they may not gain the necessary breadth of experience, it can be seen why they sometimes opt for the Commercial Property route. To do this, they must of course be comfortable with the Valuation competency to level 3.

As an example, candidates working for regeneration agencies often find that the

Commercial Property route works well. They may not be actively involved in planning applications, rather instead, land assembly, compulsory purchase, etc., and also some property marketing work (as per the competency of Purchase, Disposal and Leasing). Here, optional competencies may, for example, include Economic Development (level 3), Development Appraisals (level 3), Planning (level 2), and Purchase, Disposal and Leasing (level 2). Compulsory purchase would not have to be added as a fifth optional competency, but as experience in capital valuations may be limited (as asset/Red Book valuations), and as the Development Appraisals competency could incorporate both development valuations and development appraisals, the core Valuation competency could include compulsory purchase valuations. It would however be important that the candidate was involved in the valuation and negotiation of compensation settlements, as opposed to involvement only in surrounding legal aspects of compulsory purchase. These examples are a good illustration of the importance of candidates accurately considering competency selection having regard to their circumstances, and, within reasonable tolerances, making the APC work best for them.

The optional competencies could include any mandatory competencies which are to be taken to a higher level than the minimum level as a mandatory competency. The mandatory competency of Conflict Avoidance, Management and Dispute Resolution Procedures, for example, only needs to be level 1 as a mandatory competency, but could be taken to level 2 or 3 in conjunction with Landlord and Tenant if candidates are involved in arbitration and independent expert processes at rent review, and court and PACT processes at lease renewal.

Candidates need to avoid the temptation to select optional competencies from the full list which appear to provide an easier route through to APC success. This refers to the softer skills such as Teamworking, and Oral Communication. Consultancy Skills is primarily a softer skill, which, while inappropriate for most candidates to take, may be suitable for candidates involved in strategic/advisory functions in connection with real estate strategies/corporate estate management/ management consultancy work or strategic planning, and the like.

It would also be inappropriate for a candidate in the Commercial Property route to take Law to level 3, rather than retain this as a mandatory competency to level 1, unless particular experience was being gained in law (as opposed to law underpinning the usual day-to-day work). However, the Planning and Development route includes Law as a core competency to level 3 – possibly an example of the occasional minor anomaly in the system.

As part of final assessment, assessors will consider the appropriateness of candidates' competency selection among a range of other criteria. Candidates taking inappropriate and/or conveniently easy competencies may be judged by assessors as not having gained the necessary experience. Prior to the current post-August 2002 system, candidates selected optional competencies from a prescribed list for each APC route (and although others could be brought in, it is was uncommon for candidates to do so). It was not often that inappropriate competencies were selected. Under the post-August 2002 system, it is candidates' ability to select optional competencies from a list of over 100 competencies covering all surveying disciplines/APC routes which has increased the possibility of inappropriate competency selection. An example would be to select the competency of Marketing as well as Purchase, Disposal and Leasing, but without actually being involved in the marketing of products, services, businesses, etc., in accordance with the Marketing competency, and only being involved in lettings, sales, etc.

It is important that candidates scrutinise the RICS guides properly, and establish the work activities which may relate to a particular competency. One common mistake, for example, is candidates recording experience under the competency of Inspection because they have been on site. This can lead to an unnecessarily high number of days' experience recorded under Inspection, and too few in Valuation, for example, to which the inspection may have related. Under the pre-August 2002 system, some general practice candidates selected the competency of Property Development and Acquisition because of involvement in property development work, but without examining the RICS guides properly to find that this is a competency designed primarily for building surveyors, and not on the list for general practice surveyors). Candidates finding out about such mistakes only at the APC interview will naturally come across very poorly to assessors.

It is advisable not to take more than the minimum number of competencies, nor to take competencies beyond the minimum levels (1,2,3), other than in exceptional circumstances. This ensures that candidates do not create undue learning efforts for final assessment, and are more easily able to convey their principal areas of experience to assessors.

It is also important for candidates to be aware of issues beyond their competency headings. A general practice surveyor, for example, may not have the Local Taxation/Assessment (rating) competency, but liability for business rates will be a consideration in many areas of work, and candidates must be familiar with key elements of the rating system. Similarly, the competency of Planning may not be undertaken, but planning, including scope for alternative uses, will relate to many areas of work. A planning and development surveyor would need an awareness of Landlord and Tenant, as some development schemes may involve securing vacant possession from current tenants, with compensation payments affecting the development appraisal and possibly also, overall scheme viability.

Further details on the areas of work falling under some of the main competencies are shown in the example of the summary of progress completed at interim assessment in Chapter 4, and also in Chapter 10.

Diary recording and log book entries

Diary entries should cover core and optional competencies, bearing in mind that mandatory competencies are general skills which are represented throughout a candidate's case work.

As an example of diary entries, the extract below is provided from a candidate's diary (with locations withheld). Each entry would be allocated a competency reference. When subsequently preparing the summary of progress at interim assessment and final assessment, there is ample diary information to ensure that the reports can contain sufficient detail – and generally sell the candidate's depth of case work to assessors.

Visited Westminster City Council in Victoria to study all planning applications registered on xxxx (half day).

Meeting with Westminster City Council in order to establish the possibility of purchasing the xxxx site, and the likelihood of gaining successful planning permission (half day).

Prepared details on the xxx site and wrote a brief report on the viability of the site, and then talked through the potential purchase with the Head of Investment (1 day).

Went to inspect the office refurbishment works at xxxx. Discussed with security how best to improve this situation (half day).

Chased the remainder of the rent arrears from the December quarter in respect of xxxx offices and made a decision to bailiff two of the tenants (half day).

Spent the entire day calculating the service charge expenditure to date for xxxx. This was necessary prior to calculating a service charge budget for the forthcoming year.

Diary entries should be half-day or full-day entries. As surveyors do not, of course, undertake their day-to-day work in line with such neatly divisible days, the diary is more of a representative illustration of the work undertaken. An effective way to complete the diary can be to record work in draft, without allocating competencies, and then at the end of each month, make the diary entries. There will still be neat half-day and full-day entries, but this approach will enable some of the smaller, but important, tasks to be aggregated.

Often candidates find that an individual piece of case work covers a number of competencies, and they are unsure how to allocate the diary entry to a competency. It is important to examine the precise task involved in a piece of work, rather than consider only the broad category of work being undertaken. Candidates involved in agency/property marketing work, for example, can fall into the habit of repeatedly allocating work to the Purchase, Disposal and Leasing optional competency (Property Marketing under the previous APC system). This creates log book totals which are actually unrepresentative of the experience being gained. If analysing the case work more accurately, diary and log book entries may also cover Valuation (as rentals or sale prices will need establishing, comparable evidence sought etc.), Landlord and Tenant (because suitable lease terms will need determining/negotiating) and Planning (in the case of development sites in particular, but a shop letting could also involve a change of use, with listed building status perhaps also being relevant). If a candidate did not have the competency of Landlord and Tenant, the establishing of lease terms could remain under the competency of Purchase, Disposal and Leasing. The usual entries under the competency of Purchase, Disposal and Leasing would involve taking instructions, property inspection and measurement (perhaps shared with the core competencies of Inspection and Measurement), preparing a marketing report, formulating the marketing strategy, arranging advertising, etc. The above illustrations are a good example of how the support of supervisors and counsellors can alert candidates to wider issues in practice, and generally prompt lateral thinking – a key skill for chartered surveyors.

Diary entries should not be wasted on descriptions such as 'office administration'. Annual leave and sick leave should be stated as such, but if, for example, a candidate worked a Saturday in lieu, or had undertaken substantial over-time in the surrounding weeks, it would not be necessary to lose days simply because of the work that was not precisely timed in the allotted days. Public sector candidates working to flexi-time arrangements similarly should avoid making an entry such as 'flexi-day' on every second Friday, if nine full-days will have covered 10 days' work in order to create a day off. Professional development should be logged in the professional development record, but if a day-course has been attended, a diary entry could be 'professional development day'.

The diary is not submitted to RICS as part of final assessment submissions, but is nevertheless an important part of a candidate's APC training period. The log book cannot, of course, be completed without the diary being up to date, and will also be drawn upon for the summary of progress at interim assessment and final assessment. At final assessment, assessors could request to see a candidate's diary, although this is unlikely if submissions appear well-organised, and the APC training period has been conducted properly. RICS also arranges spot checks on candidates and employers, undertaken by the RICS Regional Training Advisers. They will expect to see all records reasonably up to date, including the diary (and also the three-monthly supervisor's and six-monthly counsellor's progress reports).

An up to date diary and log book also enables candidates to demonstrate to their supervisor and counsellor the need for wider experience.

Structured training agreement

As mentioned in Chapter 1, employers need to have a structured training agreement in place (approved by an RICS Regional Training Adviser) before their candidates can undertake the professional APC training period. Content varies between large practices with many recruits per year, and smaller concerns taking on a graduate only occasionally – with some practices using the document as a working framework for the implementation of their APC training programme.

The structured training agreement provides general information about the employer and the nature of their business, and specific information on the experience which is available, and how this relates to APC routes and competencies.

An example of a structured training agreement prepared in respect of a mid-size London practice, Saxon Law, for a specific APC candidate, following the RICS template model is set out below (with some information being withheld).

Contents

Employer's statement

Brief description of business to include:

- Number of offices.
- Number of employees.
- Organisational structure and activities.
- Training policies (fees, in-house training provision, etc.).

Competencies

- Details of competencies provided to candidates.
- Competency achievement planner.

Candidate's statement

- Candidate statement.

Appendices

- Competency achievement planner.
- Monitoring table.

Employer's statement

The company is a practice of general practice Chartered Surveyors. Our main area of work is central London offices. We employ 40 staff in one office located in the UK only. We are specialists in offices and undertake West End and provincial office agency, Rent Reviews, Lease Renewals, Management, Investment agency and associated valuations work.

Commitment to training

The Saxon Law Partnership was formed in 1985 and was established by two senior staff who had previously worked at a large practice. The firm, in its formative years, recruited staff of a senior level whilst the practice remained relatively small. However, with the expansion of the practice to a total of now some 40 people, and covering a much broader and wider range of services to clients, the firm has become fully committed to employing a greater cross section of staff in terms of age and experience. With this change in the topography of the firm, the Saxon Law Partnership has seen that investing in young trainee surveyors is vital to the progression of the firm. It is also able to offer positions for young trainees in a dynamic environment, offering hands-on experience in the work undertaken, together with the ability to work with highly motivated and qualified individuals on a one-to-one basis, and within teams.

We believe that the environment that we can offer a trainee surveyor gives both the breadth of experience required by the APC, and also a commercially exciting environment within which to learn the many skills which a surveyor will need in fulfilling his day-to-day role.

Commitment to the APC

As part of the growing need for younger members of staff, Saxon Law has become committed to providing the necessary training and experience required by the candidates for the RICS APC. We are fully aware of the guidelines that have been set out and are confident that we can provide the correct range of experience, quality of work and high level of involvement needed by a candidate in order to fulfil their requirements under the APC. We have put in place a structure of reporting and counselling to ensure that our candidates achieve the required levels, not only for the APC but also for the firm as a whole. It is important, as part of the training, that an APC candidate is able to control and balance the commercial needs of the firm and the required levels of training required by the APC. We believe that this is a very important part of both the professional and the commercial training that a young surveyor requires in the ever changing and evolving property market.

We are meeting regularly with our candidates to ensure that they are achieving the required levels of competence within the competency areas chosen, but just as importantly feel they are adding to the existing teams within which they are working. We consider it of vital importance that candidates not only obtain the correct experience, but also thrive within a busy working environment and enjoy their day-to-day working environment in order to facilitate the best results for all parties.

Employer's policies

Saxon Law Partnership, as a prerequisite, undertakes to pay all related fees to RICS for all employees, including subscriptions and APC fees, etc. Also, in addition to this, the firm pays for any additional training required by any employee, including areas such as CPD and the general furthering of business skills.

Competencies

The following is a brief description of the work in which the candidate will be involved in the various departments, together with an indication of the core and optional competencies covered by such work.

Investment

Valuations – assisting with valuations on buildings in London and the South East (*Valuation*).

Report writing – Various investments (*Mandatory*).

Viewing – Conducting viewings with prospective purchasers of various investments (*Various, and Inspection*).

Capital allowances – Part of a two-man team assisting claims for capital allowances (*Valuation*).

Measuring – Measuring various buildings as part of a team (*Measurement*).

Requirements for small investors – Following up requirements including identifying suitable properties and submitting details for consideration.

A new residential development considered for purchase by a client – investigating the planning with Westminster City Council and preparing details on the property.

Inspection – Part of the inspection team for a large portfolio, under consideration for purchase by one of the firm's clients.

Management

Day-to-day management of a portfolio of 80 buildings.

Setting up new tenants' leases onto the computer database. This will involve reading leases from cover to cover, writing lease summaries and completing computer-input forms for the management database.

Writing reports from inspections on a weekly basis.

Negotiating with tenants to collect rent and service charge arrears.

Budgeting nine service charges for multi-let office buildings for the year end 25th March 2003–24th March 2004.

Overseeing applications for assignment, subletting and alterations on several leases.

Setting up and re-tendering of maintenance contracts.

Liaising with tenants and landlords, project managers and contractors for the internal refurbishment of the common parts of a multi-let office building totalling 25,000 sq ft.

West End Agency

Dealing with the day-to-day letting and acquisition of West End office premises. This will include preparation of reports and schedules to include details of current tenant requirements, details of competing buildings, details of current availability, etc.

Undertaking inspections of buildings, either on behalf of landlords for letting/sale purposes or on behalf of tenants for acquisitions by lease/freehold purchase.

Undertaking negotiations on behalf of either landlords or tenants on office properties.

Assisting two partners within the department and working on projects with them, including development agency and undertaking residual valuation and analysis etc.

Rent Review/Lease Renewal

Inspection – Establishing the net floor area by on-site measurement. Referencing properties and paying particular attention to the specification and usability factor, and any improvements which are to be disregarded on review.

Lease reading – Identifying from a detailed reading of leases and other documentation any legal factors that may affect the rental value of the property, and the basis of agreeing the reviewed rent.

Evidence – Compiling schedules of comparable evidence of rentals agreed for similar properties.

Report writing – Learning to write reports with valuations and recommendations for strategies in order to achieve the best possible settlement.

Negotiation – Learning how to negotiate effectively, and assist team members in negotiations.

Candidate's Statement

The objectives of the Assessment of Professional Competence, as set out in the APC guides, are wholly appropriate, and I will follow them to the best of my ability. I am committed to work both on behalf of the firm, and also in order to achieve RICS competence.

I feel that qualification is necessary if I am to fulfil my ambition and follow a career within this industry. I will endeavour to keep my diary and log book up to date and as detailed as possible, and to arrange periodic reviews with both my supervisor and counsellor, including at interim assessment. I fully intend to meet the professional development hours required as part of the APC, and I shall apply for my final assessment as soon as the required competencies have been met and signed off by my supervisor and counsellor.

As an example of some of the elements included in the structured training agreement of a large firm, extracts from a large practice's documentation are as below.

... (following competency descriptions etc.) For example, the practice's general practice candidates in the Commercial Property/Commercial Real Estate Practice route, will be able to gain experience in their core competencies of Valuation (to level 3), Inspection (level 2) and Measurement (level 2).

Optional competencies commonly taken by the practice's general practice APC candidates, and in which level 3 is achievable, are Landlord and Tenant, Purchase, Disposal and Leasing (property marketing/agency), Real Estate Management, Local Taxation/Assessment (rating), Development Appraisals and Planning.

Optional competencies less commonly taken by the practice's general practice surveyors are Asset and Investment Management, Compulsory Acquisition and Compensation, Conflict Avoidance, Management and Dispute Resolution Procedures, Corporate Real Estate Management, Corporate Recovery and Insolvency, Economic Development, Insurance and Risk Management, Real Estate Finance and Funding and Strategic Real Estate Consultancy. Experience may sometimes be achieved in other competencies.

It has been considered inappropriate at the practice to outline the detailed experience available to candidates in this structured training document, as it is heavily dependent on the office and business unit in which they work. Instead, information is provided to candidates through the graduate recruitment process, and as part of APC induction training taking place on both a group and individual basis.

Training support

In addition to the day-to-day support provided by managers (including APC supervisors and counsellors), the practice supports candidates by operating an in-house APC training programme, and by providing suitable professional development/CPD training.

On joining the practice, graduates receive a detailed induction in respect of the work of the practice, and their graduate training programme.

APC candidates are given assistance in respect of competency selection, having regard to the work likely to be undertaken over their training period. This includes graduates who commenced the APC with a previous employer, and may have to change competencies to reflect the new areas of work likely to be undertaken at the practice.

A graduate rotation system ensures that candidates are able to achieve a wide range of knowledge in addition to the relatively in-depth experience which may be gained by working in the practice's business units. In some situations, candidates will be involved in other business units' case work without formally being seconded. This will particularly be the case in smaller offices where individual surveyors at graduate level cover a wider range of work, often across several business units.

Appendix 4 is the 'Candidate summary form of APC training programme'. This helps the practice's managers, APC supervisors and counsellors, and also the company's training personnel ensure that APC candidates are gaining the requisite work experience. RICS 'competency achievement planner' also helps candidates to monitor their progress.

Candidates are supported financially regarding RICS membership and their assessment and training fees, and are allowed time to prepare written submissions, attend training etc. in work time. At final assessment for the APC, up to 10 days' study leave is available, and a further day's leave is granted in order to attend the interview.

Professional development

As part of the graduate training programme, professional development/CPD events are run specifically for APC candidates. This is part of a wider programme at the practice which helps ensure that individual learning and personal development is accelerated by enabling graduates to build on their work experience with suitable subject based training.

A summary of the areas covered as part of general practice candidates' training is set out in Appendix 2, 'Summary of professional development/CPD training programme'. For candidates in other routes, different arrangements, such as attendance at external events, often apply, owing to the lower candidate numbers not making in-house group sessions viable.

Progress reports and the record of progress

Supervisors complete a three-monthly progress report, and counsellors a six-monthly progress report. These are not submitted to RICS as part of final assessment, but as with the diary mentioned above, can be requested by assessors, or required as part of RICS spot-checks at any stage of the APC training period.

Template formats are as per the RICS guides, also available via RICS website.

The following is an extract from a candidate's three-monthly supervisor's report, including comments from the candidate. Some information is excluded.

Observations on training to date, experience gained and ability of candidate

During this period James has been working within Saxon Law's Investment Department. As part of his job he joined the department whilst the firm was involved in the acquisition of a substantial West End office building on behalf of private clients. This was an extremely detailed acquisition which involved James in a number of issues, including the measurement of the building, which is in the region of 350,000 sq ft, and valuations for a range of purposes.

James was involved in a number of asset management issues and also attended legal meetings, together with specialist surveying meetings (including specialist surveys such as contamination, mechanical and electrical and structural works). He also liaised with tax consultants and capital allowance consultants. James was very much an integral part of the team working on this project, and his contribution was highly valued by the team, and also by Saxon Law as a whole. He undertook all his tasks with great enthusiasm and importantly with great competence, and was able to undertake certain tasks with minimum supervision because of the confidence the team had in him.

In addition to this major acquisition, James was also involved in two bank valuations of smaller buildings, including one in Farm Street on the basis of a residual valuation. He valued the conversion of an existing refuse site to an office scheme and, secondly, a building in Queensgate Terrace for a conversion from an existing hotel to a residential scheme. Both of these valuations included site inspections, existing use valuations and residual valuations.

James has also been involved in undertaking searches in order to fulfil a requirement for a number of small investors of the firm (which includes liaising with both property owners and agents in sourcing potential opportunities for these clients). This has also included undertaking inspections with clients and opening negotiations on their behalf. He has also been involved in the sale of an investment building in Curzon Street, undertaking numerous viewings with interested parties, and entering into negotiations with some of the parties.

James was also involved in the sale of a small central London portfolio of three buildings which again involved site inspection, measurement, advice to the client, collating investment sale details, undertaking inspections, entering into negotiations and reporting to the existing owners on progress and marketing. Most importantly, he should be applauded for his work undertaken on this particular project, as he handled the day-to-day running of this instruction whilst the partner leading the project was on a four week sabbatical over the early part of 2003.

I am delighted with James' progress and his ability to grasp the various concepts and technical issues that he has been faced with. In addition, his ability to prioritise and manage his workload whilst still maintaining the various requirements of the APC has been outstanding.

Candidate's comments

My first three months have been spent in the Investment Department working alongside my Counsellor, Andrew Lax, who is Head of Department. I have been very fortunate to work on some large scale projects during these three months, and the opportunities that have come to me have been both diverse and challenging.

I found myself involved in valuations, measuring, capital allowance calculations and working on some small freehold acquisition and disposal work myself. I have been very fortunate in the amount of time that other members of the team have afforded me, and I have had outstanding advice offered to me. I feel that I now have a much greater insight into the profession and have had some wonderful opportunities to take in some very detailed information. I believe that I have grasped the main theories and concepts behind what drives good investments, and my first three months have given me a hunger to better my knowledge and understanding.

The record of progress is the schedule of candidates' competencies signed off by supervisor and counsellor. RICS guides also, however, refer to the record of progress as incorporating progress reports and interim and final assessment submissions. In *How to Pass the APC*, the record of progress refers to the schedule of competencies.

Professional development

In addition to the work experience being gained, candidates undertake professional development, which basically must comprise a minimum of 48 hours per year. The RICS guides provide further detail, including on how areas of professional development can be categorised. As an illustration of how a good quality programme of professional development can be achieved, an extract from one of the larger private practice's structured training agreement is set out below. For smaller firms without such centrally co-ordinated training programmes, APC candidates can become involved in arranging CPD (the equivalent to professional development for qualified surveyors) for their colleagues.

The professional development/CPD opportunities available to the practice's graduates include the following.

- In-house APC workshops covering the main subject areas such as valuation, investment, marketing/agency and landlord and tenant, run at the above mentioned regional offices.
- In-house workshops covering the more specialist competency areas taken by fewer candidates, such as rating, insolvency and planning and development, run at one single venue on a national basis.
- Attendance at external evening seminars, and half-day and full-day courses.
- Undertaking a range of private study initiatives. This will include pre- and post-event material as part of the in-house subject based APC workshops.

- Occasionally preparing presentations/seminars for subject based APC workshop sessions.
- Assisting with the research and writing of articles for professional journals, the practice's bulletins and other promotional material.
- IT, presentation, interpersonal, negotiation, business etc. training as part of programmes for all surveyors.
- Involvement in subject based CPD events arranged for qualified surveyors.

The overall training support provided should see candidates substantially exceed the minimum 48 hours per year of professional development/CPD required by RICS as part of the APC.

In-house workshop style sessions will combine the formal delivery of technical information with group discussions regarding cases in which candidates are involved. It is this practical style of training which helps candidates build a knowledge and understanding that equips them for the APC interview – as well as directly helping with case work.

The practice's qualified surveyors will contribute to some of the APC workshops, such as by providing a short seminar, or by discussing current issues within their specialist field.

It is feasible to undertake professional development/CPD training sessions specifically for general practice APC candidates on a regional basis owing to the relatively large number of candidates.

Building surveyors and planners will still be involved in some of these events, but in view of their smaller representation, their training will generally involve different activities.

Further information on professional development/CPD is included in Chapter 13.

Chapter 3

Key Input from Supervisor and Counsellor

The supervisor and counsellor have an important role to play in a number of areas. This brief chapter summarises the key aspects for the supervisor's and counsellor's convenience, cross-referenced to the relevant sections within *How to Pass the APC*. These are:

- Gaining a basic familiarity with the APC. (See Chapter 1, page 1.)
- On behalf of their employer, ensuring that a structured training agreement is in place. (See Chapter 2, page 16.)
- Ensuring that candidates register to commence the APC as soon as they begin employment. Delays can lead to the final assessment interview being put back. (See Chapter 2, p7.)
- Helping candidates understand aspects of the APC. It is not the responsibility of supervisors and counsellors to learn the minutiae of RICS APC requirements, to then be able to instruct candidates, who should be equally as able to establish the requirements themselves. Particular support will, however, be required by candidates in determining the type of work in practice falling in each competency area, and also how the work experience provided by the employer fits in with competencies. (See Chapter 2, page 8.)
- Related to the above, working with candidates on competency selection. (See Chapter 2, page 9.)
- Working with candidates on the completion of the competency achievement planner. This is part of the initial registration requirements, and maps out candidates' anticipated stages of achievement against each competency. Although, in practice, this may not be established with any accuracy, the process helps candidates and their supervisor and counsellor focus on the APC requirements. (See Chapter 2, page 9.)
- Ensuring that candidates are gaining the right work experience in relation to competency requirements.
- Encouraging candidates to take a pro-active approach to learning by thinking about their case work in due depth, and undertaking related study. Examples include scrutinising the Red Book when undertaking asset valuations, and studying the *RICS Code of Measuring Practice* when measuring properties.
- Encouraging candidates to take a committed approach to professional development – APC candidates' equivalent to qualified surveyors' CPD. (See Chapter 2, page 23.)
- Ensuring that the diary, log book and professional development records are up to date.
- Undertaking regular review meetings with the candidate.
- Completing three-monthly supervisor's reports and six-monthly counsellor's reports, and completing supervisor's and counsellor's reports as part of interim assessment and final assessment requirements. (See Chapter 2, page 21.)

- Completing candidates' record of progress as they attain the required competency levels. (See Chapter 2, page 21.)
- Assisting candidates regarding interim assessment, including ensuring that this is completed on time. (See Chapter 4, page 27.)
- Assisting candidates with final assessment submissions, in particular the critical analysis. (See Chapter 5 on final assessment submissions, Chapter 6 on the critical analysis, and Chapter 7, which provides an example of a critical analysis).
- Ensuring that candidates meet the required competency levels at final assessment before being signed off as competent to sit the interview (or, for practical purposes, being satisfied that they will meet the necessary levels following a further two to three months of work experience and surrounding learning – noting also that competencies can be signed off just prior to the interview, as opposed to at the point when candidates forward their submissions to RICS).
- Providing mock interviews near to the final assessment interview. Mock interviews could be undertaken at additional points in the training period, and could link with performance appraisal functions. (See Chapter 10 on what candidates need to know in their competency areas, and Chapter 11 which provides examples of interview questions.) It is important to note the competency and experience based nature of the assessment process, and ask questions regarding candidates' case experience – rather than only technical questions which can often be the case when managers provide mock interview without having had experience of the APC. Supervisors and counsellors will also assist with candidates' preparation of their 10 minute presentation as part of the interview.

As stated in the preface on page vii:

> The national pass rate for general practice surveyors taking the final assessment interview has been around 65% over recent years (which after allowing for re-sits, equates to an approximate 50% first time pass rate). This is low, considering that one of the elements of the APC is that employers (as supervisor and counsellor) are required to declare that candidates have reached the necessary levels before putting them forward for final assessment.
>
> Reasons for the number of unsuccessful candidates include a lack of awareness as to what exactly needs to be understood within competency areas, supervisors and counsellors signing off candidates despite not having met the required levels, and candidates being prepared to take a speculative gamble on APC success – consoling themselves that failure would still be good experience for further attempts. A related factor is candidates often focusing on two months of revision, rather than at least two years of concerted learning.

If candidates fall considerably short of the required competency levels, RICS may visit the supervisor and counsellor to provide the necessary support in observing the RICS APC requirements.

Chapter 4

Interim Assessment

Interim assessment provides an opportunity for the candidate and their employer to review their first year of training, and plan their continued combination of work experience and surrounding professional development study for the period leading to final assessment.

Timescales

Interim assessment takes place after 12 months' experience have been gained, and should be completed within one month thereafter (i.e. before 13 months' experience have been gained). The start date for candidates' training period is their 'registration' date, as mentioned in Chapter 2, Starting the APC.

If interim assessment is not completed on time, this can cause a candidate's final assessment interview to be put back, as RICS requires 12 months' post-interim experience to be achieved before candidates are eligible for final assessment. However, candidates need to be aware of RICS precise rules, which are more detailed than the statements included in the RICS guides – such as the above 12 months' criteria being up to the end of the assessment period (30 April and 31 October for general practice surveyors), rather than the actual date at which the final assessment interview is sat.

Components of interim assessment

Interim assessment is completed by the candidate together with their supervisor and counsellor. It is not submitted to RICS at the interim stage, but will form part of final assessment submissions. The assessment comprises:

- Diary – complete for the first 12 months.
- Log book – up to date, stating total number of days' experience in each core and optional competency over the first 12 months.
- Professional development record – up to date, and containing a minimum of 48 hours (of suitably qualifying activities).
- Summary of progress – a 2,000 word report outlining the experience gained over the first 12 months.
- Forward plan – a 1,000 word report outlining the experience expected over the period to final assessment, and how the necessary competency levels will be reached.
- Record of progress – candidates being signed off by their supervisor and counsellor as having met competency requirements/levels.
- Supervisor's and counsellor's report – comments by the supervisor and counsellor on the candidate's progress.

As always, candidates need to check RICS precise requirements from the relevant guides.

RICS spot checks

The diary is not submitted to RICS as part of final assessment submissions (nor are the three-monthly and six-monthly reports prepared by the supervisor and counsellor). However, although unlikely for candidates with well-prepared submissions which demonstrate a proper and organised approach to their professional training period, assessors may request the diary and other records at final assessment. Also, and more likely, RICS undertakes spot-checks through the Regional Training Advisers on individual candidates, and on employers' implementation of the APC.

RICS template reporting formats

The summary of progress and forward plan are completed on templates provided by RICS – as shown in the RICS APC guides, and as available to download from the RICS website. A template is also available for the supervisor's and counsellor's report at the interim stage.

For the summary of progress, candidates outline their experience against their competency headings, which comprise mandatory competencies, core competencies and optional competencies. A final section of professional development is also completed.

The RICS template does not provide a heading of introduction, but it is worthwhile candidates creating such a heading, either at the start of the template, or on an additional cover page. The forward plan can be completed similarly, to include an update, for example, on a change of employer or other circumstances. (Candidates having undertaken interim assessment prior to the current system will have listed 'common' competencies, which are equivalent to the new mandatory competencies.)

An example of an interim assessment is set out below, but first, the following points need to be considered in respect of competencies:

- Competencies are not often signed off to level 3 after only 12 months' experience. Exceptions include candidates working primarily in a particular area – such as a valuation department of one of the large practices.
- The signing off of all competencies at interim (as does sometimes take place) can suggest to assessors that the employer is not operating the training programme properly.
- Flexibility needs to be preserved regarding the areas of work to be undertaken during the second year or more, and which competencies will be most suitable at the final assessment stage. Although competencies can be changed during the APC training period, candidates should, for example, avoid being signed off to level 3 at the interim stage, only to later wish to reduce this to level 2.
- It may be the case at the interim stage that candidates do not yet know what their full range of competencies may be. This just requires a suitable

explanation in the summary of progress and forward plan – such as being part of a rotation scheme throughout the departments of a major practice. It is not a problem that work may be unbalanced throughout the training period.

The impact on final assessment

Comments were made in preceding chapters regarding the need to think about how all stages of the APC can have an influence on final assessment, and therefore require thought and assessment on an ongoing basis. Examples include not selecting more than the minimum amount of competencies, except in special circumstances such as a change of department, employer or caseload which makes it unavoidable. Similarly, competencies taken to level 2 should not be signed off higher than is necessary.

Where too many competencies are taken, and/or competency levels are unnecessarily high, consequences include increased learning efforts at final assessment, assessors' expectations being raised, assessors not be able to see which competencies represent a candidate's strengths, and tougher questioning being received in level 3 competencies which could otherwise be to level 2.

If candidates wish to express any additional experience they have gained, rather than creating additional competencies, a section on the lines of 'Other experience gained' could be added to the template just prior to the section on professional development.

Any additional competencies taken only to level 1 should not however cause difficulties at the final assessment interview.

Assessors' views on interim assessment records

The various aspects of interim assessment will need to be signed off by the candidate, supervisor and counsellor. Completion and dating some way beyond the due date for completion risks creating a bad initial impression to assessors, as well as possibly causing RICS to put candidates back to a later assessment date owing to ineligibility for the reasons mentioned above.

Also, as indicated in Chapter 5 relating to final assessment submissions, assessors will examine how the final summary of progress interrelates with the comments made in the forward plan at the interim assessment stage.

Example of interim assessment

In order to demonstrate how an interim assessment is completed, set out below is a typical example from guest contributor James Turner, an APC candidate with Saxon Law, specialising in office agency and related investment and asset management strategy in London's West End. Some aspects have been disguised for confidentiality. Also, whereas the actual interim assessment would be completed on the RICS templates, only the principal text is repeated.

The word count of the illustrative summary of progress considerably exceeds the usual 2,000 words, but the example is nevertheless included here, as it also highlights some of the areas of experience candidates will need to gain. It is also worth noting that the 'training planned' aspect is relatively detailed. Candidates sometimes add comments into the 'training planned' section when writing the summary of progress, and then either repeat themselves or think of relatively limited further comments in the forward plan. It is therefore preferable to complete the training planned section of the summary of progress once the forward plan has been finalised.

It is important to highlight that although examples provide helpful support for other candidates on the structure and broad content of the interim assessment, the best interim assessments are driven by candidates' own work experience and circumstances. As with all examples of reports, candidates trying to mirror another candidate's work too closely tend to produce a poor report, skewed to the example rather than accurately reflecting their own situation.

It is important that candidates convey the due depth of their work to assessors, as well as the breadth of experience gained. This can be achieved by reference to case examples, including, for example, key negotiating points, factors affecting value, and legislation and case law which may have been drawn on.

Interim assessment – summary of progress

Introduction

I graduated from Oxford Brookes University in June 2002 with a BSc (Hons) accreditation in Real Estate Management.

The course focused on Landlord and Tenant, Corporate Real Estate, Real Estate Management, Valuation and Appraisals, Project Management and Computer Technology.

In September 2002 I started work at Saxon Law, a niche property and development consultancy practice. My first placement was in the Investment Department where I spent my first three months before joining the Management Department. Changing departments has given me the best possible experience for the APC and for the development of my career.

After completing six months within our Management Department I moved to the West End Office Leasing Department.

Mandatory competencies

Code of Conduct, professional practice and bye-laws

By working in a smaller firm I have been given more responsibility, certainly with regards to dealing with clients face to face. In the Management Department especially, a clear understanding of a particular client's needs, and how these tie into RICS rules are key.

I have learnt the importance of ethics and the relevance of applying the correct legislation in order to fulfil client's requests, as part of providing a professional service. External reading has taught me the power of negligence and how it can be the downfall of not only an individual but also the individual's employer. I have, at various times throughout the year, referred to the RICS Red Book on different issues, as well as asking colleagues for their views on matters.

(Training planned) I hope over the next year to further improve my knowledge of professional practice and bye-laws. I aim to do this by taking heed of my colleagues' advice and by asking as many questions as possible. I plan to spend more time becoming familiar with the RICS Red Book, with specific attention being paid to the bye-laws, along with attending in-house courses.

Conflict avoidance, management and dispute resolution procedures

I have become more cautious as experience has developed. This is as a result of mandatory checks being carried out by each employee upon the instruction of any new work. I am confident that I can avoid conflicts and understand the importance of undertaking searches for conflicts. My dispute resolution procedures knowledge is weak at present. However this is something I will be looking to improve throughout the next 12 months.

(Training planned) Dispute Resolution Procedures is something I aim to learn more about both through the workplace and through my colleagues. *Estates Gazette* and a variety of other trade publications, along with any related courses, will be valuable learning references.

Collection, retrieval and analysis of information and data

This competency has contributed significantly to my experience so far. In the Investment Department I spent a great amount of time using both Kel Delta and Sigma in order to analyse information and data to calculate site and development values. I gained further experience within the West End leasing department. A lot of my work involves carrying out searches for either deals done or properties available within certain areas on EGi, Focus or Streets Ahead. Using packages such as Kel has benefited me hugely. I now have a good insight into how the internet and search databases play a key role to the surveying industry on a day-to-day basis. I have also attended various advanced Excel spreadsheet courses in order to maximise the ways in which I am able to analyse data.

(Training planned) I will attend further internal advanced IT training courses. I hope to expand extend my knowledge and use of programs such as Kel, in order to work to optimum ability.

Customer care

Whilst spending six months in the Management Department it became clear to me, especially when handling unsatisfied tenants, that customer care is a vital aspect in property. As a management surveyor, you often take sole responsibility. The West End Leasing Department has strengthened my ability to handle clients as much of the work is speculative, and trying to tie a client to a deal can be made or broken by good customer care.

(Training planned) To continue to value the importance of this competency, I think it is necessary to attend in-house courses relating to customer care.

Customer care can also be improved through increased confidence, which I hope I will gain by further experience.

Environmental awareness

The majority of my experience relating to environmental awareness came through my time in the Management Department; however this was limited. I worked on asbestos issues at the time when a search was carried out across the Berkeley Square Estate. I attended meetings with specialist surveyors in contamination and they taught me just how stringent they have to be over environmental issues.

(Training planned) This is an area of my training that needs to be improved. I intend to read many more articles in *Estates Gazette* and *Property Week* on this topic. I also intend to attend a CPD lecture on this subject.

Law

This competency is one that was taught heavily at university, and through this sound theoretical knowledge I have been able to combine my previous learning with the legal requirements that now govern my work on a day-to-day basis. By attending both pre-lease and lease meetings, I have gained a better understanding and a more practical, rather than theoretical, insight into the legal process. This involved meetings with both my clients and their solicitors, and the opposite side's clients and solicitors with specific relation to pre-lets. I have a good understanding of the Landlord and Tenant Act 1954 with regards to the tenant's right to renew their lease, and the effect that this has upon new leases with regards to rental values and the security of tenure for a tenant. Whilst in the Leasing Department I also learnt about the Property Misdescriptions Act (1991) and the Estate Agents Act (1979).

(Training planned) It is vitally important that over the next 12 months I continue to increase my legal reading, paying attention to those points that affect my work directly. It will also be important to attend more law based lectures in order to keep up to date with any changes in the law, along with other laws that may not have crossed my practical experience as yet.

Health and safety

The majority of my health and safety experience has been as a result of various site inspections of properties under my management. During this period I have learnt the importance of matters such as wearing hard hats whilst on building sites along with following good practice and routine fire evacuation procedures. I have also learnt the importance of having a health and safety specialist responsible for members of the various buildings under my management.

(Training planned) It is important for me to read the regulations with regards to health and safety over the next 12 months along with gaining an improved understanding of the law relating to this matter. This may involve either attending a lecture on the topic or continued reading of various publications such as *Estates Gazette* or *Property Week*.

Information technology

Throughout the last 12 months I have been using Microsoft Word, Excel, Outlook and PowerPoint on a daily basis. I have also been trained on various search engines including Focus, EGi, Promap, Kel Sigma, Kel Delta and APB. Following on from my IT training at university, this has furthered my knowledge and use of these various systems to the point that I am competent on all of these programs. Whilst in the Management Department I learnt a great deal about the accounting databases. Whilst in the Investment Department I learnt about the valuation packages and now in the Leasing Department I have learnt a great deal about the search engines.

(Training planned) I am to continue to use these programs on a regular basis and establish a greater understanding of the ways I can get maximum use from them.

Oral communication

My interpersonal skills have improved as a result of a great deal of experience being gained in approaching unknown clients, unknown colleagues and other agents. This is especially relevant at building launches and various building open days. It is important to communicate with other agents and clients effectively and to be approachable and communicative to the best of my ability.

(Training planned) I plan to attend internal presentations skills courses.

Self management

This competency is of vital importance in the Leasing Department as the majority of time is spent undertaking viewings, and attending lease meetings or building launches. One has to be highly organised in order to manage time efficiently and put every minute of the day to good use.

(Training planned) I hope to improve my self management skills by utilising my time to the best of my ability and paying particular attention to the time spent in the office.

Team working

Having been part of three separate teams during my first 12 months I have learned the importance of team work. It was the efforts made by my colleagues that allowed me to learn so much and gave me the confidence to carry out my work to the best of my ability.

(Training planned) The bond between the team grows with time, as you learn to trust people and get to know each other better. With this in mind I aim to build on the relationships I have already made and make new recruits to the team welcome.

Written/graphic communication

Whilst in the Leasing Department the majority of my work has been to undertake various schedules and reports. If, for example, a client requires 1,000 sq ft of office space in Mayfair, it is my role to put a report together on all the available offices within that area to suit their requirement. It is also my role within the department to put together schedules of space that is available in the market. Written and graphic skills have been a fundamental part of my role so far.

(Training planned) I hope to work on reports relating to larger requirements. I would also like to improve the availability schedules in order to benefit the whole department.

Negotiating skills

My main practical experience of negotiating has come whilst working in the Leasing Department, where I worked on a number of big disposal and acquisition projects including a number of pre-lets.

(Training planned) I am confident that my negotiating skills will improve and I hope to learn from the partners that I shadow whilst negotiating larger deals.

Business management

At Saxon Law, I have learnt a great deal about business management and what makes a business attractive. It is important that structures are in place and that the company and its employees all know their end goal. An aspect of Business Management that I have particularly enjoyed learning about is how much of an impact the world economy can have on all businesses.

(Training planned) Over the next year I will keep abreast of all current affairs, paying particular attention to changes in relevant legislation. This includes reading daily newspapers and property journals.

Core competencies

Valuation

During my period in Investment I completed two bank valuations. One of these was a property on Queens Gate Terrace, comprising three terraced houses knocked into one, currently being used as a hotel. The bank asked us to value the property on a loan security basis for residential conversion. This involved inspecting the property, consulting local residential experts as to the completed unit price and collating and valuing comparable evidence. Whilst I was undertaking this valuation I also completed statutory enquiries in order to determine the other factors that may affect the value of the property. The second valuation I undertook was on behalf of a prospective purchasing client of a refuse dump in Farm Street W1 that was to be redeveloped into a hotel. This involved undertaking an extensive search in order to gain comparable evidence, inspecting the property and establishing values of hotels within the area. I also went to the Victoria Planning Office to read planning applications in the local area to establish whether a change of use and re-building would be a viable option. Following this I carried out detailed valuations and did the first drafts of the report. My knowledge of the RICS Red Book has improved considerably since University. However, a great deal more experience in Red Book valuations is needed and I would hope that over the next 12 months this important experience will be gained.

I have also attended various CPD lectures and seminars relating to valuations, which I found particularly useful with regards to the APC.

(Training planned) I will continue to attend valuation related seminars and lectures, both in-house and through the RICS. I will prioritise and further familiarise myself with the RICS Red Book.

Hopefully, when I return to the Investment Team in June, I will be able to carry out my RICS Red Book valuations along with other valuations in order to enhance my knowledge on the topic.

Inspection

Property inspection has formed a fundamental part of my training over the last 12 months. In the Investment Department I found myself inspecting a number of properties from both a purchase and disposal point of view. When in Management, inspection helped me familiarise myself with the various tenants, and the issues that related to their property and their demise. In Agency, it is incredibly important, as one is inspecting property nearly every day either on behalf of a purchaser, looking to inspect from an enquiring point of view or from a disposal aspect on behalf of a client. Whilst in the Leasing Department I have also carried out a great deal of measurement which requires a close inspection of the property. It is important before one inspects the property to have plans, details on the surrounding area and any other information I may feel will maximise the outcome of my inspection. It is good practice to know before inspecting the property what type of property one may be inspecting, the condition it is in, and any potential defects that the property may have. An inspection is vital for tasks such as measuring but also incredibly important for marketing, valuing, acquiring, disposing or whether one is simply using it for comparable evidence.

Whereas 12 months ago I would have walked around a building with no prior knowledge of it, I now feel that I have learnt that it is important to research a building prior to inspecting, in order to maximise the outcome of the inspection.

(Training planned) I plan to undertake a number of inspections over the next 12 months and hopefully my technique in inspecting property will continue to improve. I have been very focused on the West End office market over the last 12 months, and recently I have become slightly more involved in different types of unit, mainly retail. This is very beneficial to me as I continue to learn how one inspects different types of properties differently. It is also good to carry out inspections with other more senior surveyors as they may have a different approach to inspecting the property. I will also attend any future CPD courses related to property inspection.

Measurement

Whilst in the Leasing Department and to a lesser extent the Investment Department, I have learnt the importance of measuring buildings. I am regularly required to measure buildings with and without architectural plans and I feel that my measuring skills have improved a great deal throughout the last 12 months. This has mainly been due to gaining experience with more senior colleagues and also by undertaking measurements on my own using the *RICS Code of Measuring Practice* (5th ed). I now have a much greater understanding of what should or should not be included whilst measuring on gross internal or external and net internal or external bases. I feel that my attention to detail and accuracy whilst measuring has improved a great deal, and through plenty of practice I have become a great deal more competent in the ways I go about measuring more awkwardly shaped buildings. This has been very important bearing in mind that the majority of my measuring has been undertaken on the Berkeley Square Estate, which consists of predominantly period buildings with often difficult features. I have learnt the importance of good preparation and having detailed plans, a laser measure, a metre rule, a tape measure and a blue and red pen.

(Training planned) I plan to undertake further measuring exercises over the next 12 months and to learn more from my senior colleagues, helping them measure larger instructions or acquisitions. I also plan to study the *RICS Code of Measuring Practice* (5th ed) in order to further familiarise myself with the guidance and also to keep up to date with any changes relating to this.

Optional competencies

Development appraisal

My development appraisal experience came whilst in the Investment Department. This involved carrying out two residual valuations in order to calculate the alternative use values of two specific properties. This involved site inspections, establishing relevant comparable evidence, identifying and assessing the probability of planning permission, and establishing the impact of build costs and lag times. I have a better understanding of both local development plans and unitary development plans. I also gained experience of section 106 agreements and the likely impact they may have upon valuations.

(Training planned) On my return to the Investment Department I hope to undertake more development appraisals, working closely with the lead partner for central London development. I also plan to attend any relevant lectures and seminars.

Purchase disposal and leasing

I have worked either following a partner or on my own projects. I have had experience of acquiring premises, disposing of premises, surrendering leases, renewing leases, and regearing leases, as well as working on pre-lets. I have gained first hand experience of taking a client from a cold call through to completion of a new lease, and I have had first hand experience of disposing of an office suite to a client of a fellow agent. I have also learnt the importance of marketing, especially with regards to disposing of office space. Measuring and inspection have also played a large part in the purchase and disposal of leases.

(Training planned) I intend, both first hand and by following my head of department, to see more purchasing and disposal of leases. I also hope to attend any CPD lectures or seminars, and in-house training.

Real estate management

During my time in the Management Department I learnt the importance of the role of a managing agent in order to overlook the running of a building. My duties included looking after service charge properties, authorising landlord's consent, authorising invoices, overlooking contractor's works, overlooking major building works, chasing rental arrears and, to a certain extent, some asset management on behalf of the landlord in order to achieve the maximum results from their assets.

I learnt a lot about the purpose of service charges and how the service charge was apportioned. My role included calculating the exact expenditure to date and budgeting for the forthcoming year.

My time in the Management Department also helped me to establish a grasp of the relationship between the owner and the occupier of a building. The owner is always looking to maximise the capital value of a property, and the occupier is looking to rent the space at the best possible rate.

Having completed six months in the Management Department, looking after some 14 service charge properties, I feel comfortable that this experience has taken me to Level 3, and that I fully understand the role of a managing agent.

(Training planned) I will attend all relevant CPD lectures and seminars and in-house training days. I also aim to keep in touch with my colleagues from the Management Department and to keep abreast with any current issues or matters arising from the properties that were previously under my management.

Asset and investment management

To date, my practical experience on this competency has been limited to the small amount of time spent in the Management Department. However, following my move back to Investment in July, I hope that more experience will be gained. Having completed 12 months of practical experience I am now far more aware of the principles of real estate as an investment, and the reasons why real estate has proved to be a good and a bad investment in the past. Through working closely with the Berkeley Square Estate, I have learnt that the way in which one manages their property as an asset can have a huge impact on its overall value and performance.

(Training planned) I plan to develop my knowledge of this competency further upon my return to the Investment Department. However I feel it is also important to read various articles in *Estates Gazette* with regards to working property assets, along with attending any lectures and seminars or in-house training.

Professional development

Whilst at Saxon Law I predominantly relied upon software and computer training as part of my continued professional development. There are also a large number of internal lectures and seminars geared at giving excellent continuing professional development. Having attended many of these lectures and seminars ranging in topics from valuation, marketing to landlord and tenant, I have realised the importance of CPD. Over the past 12 months I have made a point of making time each week to read both *Estates Gazette* and *Property Week*.

Interim assessment – forward plan

Mandatory competencies

Code of Conduct, professional practice and bye-laws

During my first 12 months of training I have had only partial exposure to the RICS Red Book, and as a result I feel it is important that over the next 12 months I familiarise myself with the bye-laws and codes of conduct held within it. Within both the Investment Department and the West End Leasing Department, it is important that I fulfil this competency to the best of my ability, as it forms a vital role when dealing with both clients and their agents.

Conflict avoidance, management and dispute resolution procedures

My ability within this competency will greatly improve over the next 12 months through increased experience and exposure to my various roles within my department. It will also be greatly enhanced by attending any lectures or seminars with specific regard to dispute resolution, which as yet I have had little exposure to.

Collection, retrieval and analysis of information and data

My ability within this competency will improve whilst carrying out a number of searches in order to fulfil client requirements, along with collating schedules. It will also be improved by attending various IT courses, thus enhancing my depth of knowledge in programs such as Excel.

Customer care

My customer care abilities are greatly enhanced by accompanying more senior surveyors on viewings and to client meetings, and so in the next 12 months I intend to accompany senior surveyors to more of their meetings with regards to larger scale projects. This may well be strengthened further by attending any in-house courses.

Environmental awareness

I feel it is very important to attend all seminars or lectures both internally and externally over the next 12 months relating to this issue. I also intend to read *Estates Gazette*, *Property Week* and other trade publications.

Law

My practical experience of this competency has been limited over the past 12 months and I expect over the next 12 months that a great deal of time will be spent reading through all the relevant documentation relating to this. It is my aim to gain increased practical experience in this competency. I intend to participate in all lectures and seminars, both internally and externally with regards to this competency, along with reading *Estates Gazette*, *Property Week* and other trade publications.

Health and safety

Whilst in the Management Department I was up to speed on health and safety issues, it is important that over the next 12 months I keep abreast of any changes in heath and safety laws, and continue to improve my knowledge and understanding of those laws by further reading.

Information technology

I feel that over the past 12 months my knowledge of information technology has been greatly enhanced. I feel that through further practical experience and by attending further internal courses, my knowledge of programs such as Kel Delta, Kel Sigma, Excel and research programs will increase.

Oral communication

It is my intention to undertake a course in public speaking over the next twelve months and to give more input into group conversations, especially department meetings as my knowledge strengthens.

Self management

As I become more familiar with my workload I am confident that my self management will improve. I feel that through further practical experience my time organisation will improve and, as a result, the clarity of my work will be enhanced.

Team working

Having worked in two small teams I am now working in a larger team consisting of 10 people. Over the next 12 months I would expect that my team working will improve and this will be enhanced by the large team and the nature of the work that is carried out being one that relies upon a strong bond between the team.

Written/graphic communication

I expect over the next 12 months that will be entrusted to write more client reports, and therefore this competency will improve. It is also my intention to attend various in-house seminars relating to this competency.

Negotiating skills

My negotiation skills will improve as I spend the next six months in the West End Leasing Department. The majority of our work involves negotiating with other agents and with clients, and as my practical experience continues I would expect my skills to improve.

Business management

Over the next 12 months I expect this competency to improve mainly through attending lectures and seminars, both internally and externally, relating to this topic.

Core competencies

Valuation

To date my experience of this competency has been limited; however, I hope that the move back to the Investment Department will allow me to undertake more valuation work. I intend to familiarise myself in much greater detail with the RICS Red Book. Should I not get the practical experience that is necessary within this competency, I intend to get myself up to date with changes in policy or legislation through reading *Estates Gazette*, *Property Week* and other trade publications. I also intend to participate in any relevant CPD lectures and seminars, whether internally or externally, over the course of the next 12 months.

Inspection

During the next six months in Agency I intend to continue to carry out a number of inspections. These will involve measuring and valuation work and a number of viewings with clients and colleagues. It is my aim over the next 12 months to be able to inspect different use types of property, as the previous 12 months have predominantly been occupied with inspecting office premises.

Measurement

I am sure that over the next 12-month period a great deal of time will be spent measuring properties both on behalf of acquisition clients but also when working on disposals. It is also my intention to further familiarise myself with the *RICS Code of Measuring Practice* and to keep up to date with any changes to the guidelines.

Optional competencies

Development appraisals

It is my intention to work closely with the lead development partner of Saxon Law and with this in mind I hope to gain a good deal of increased exposure to development appraisals. It is also my intention to become more competent and familiarise myself further with Kel Sigma and Kel Delta. I also intend to participate in any CPD lectures and seminars relating to this competency.

Purchase disposal and leasing

Within the West End Leasing Department my knowledge and understanding and practical experience of this competency will be greatly enhanced. It is my intention to take the experienced gained over the last four months in this department further, and attempt to undertake more detailed work on my own behalf, as well as assisting in larger deals being undertaken by colleagues.

Real estate management

The majority of my real estate management experience has been achieved in my first 12 months, however, during the second 12 months I intend to keep up to date with all major changes taking place in the properties which used to be under my management. It is also my intention to be particularly well read over the industry as a whole within the next 12-month period, and therefore be up to date in any major news concerning real estate management and the relationship between both landlords and tenants.

Asset and investment management

During the next 12-month period it is planned that I will become more involved with our Investment Team, the lead partner of which is also head of the asset management role with regards to the Berkeley Square Estate. Working closely within this department I believe will give me the insight and practical experience needed for me to be up to speed with this competency. I also intend to participate in any CPD lectures and seminars in relation to this topic.

Professional development

It is my intention to continue to attend all relevant CPD lectures and seminars both internally and externally over the next 12 months. Despite achieving 54 hours of CPD throughout my first 12 months, I feel that a lot of this was devoted towards information technology, and what will be extremely beneficial over the next 12 months would be for me to attend more real estate based activities.

Interim assessment – supervisor's and counsellor's report

Supervisor's comments

Introduction

James's training to date has been structured to ensure that he has experience in line with his required training for the mandatory, core and optional competencies. His role within the Investment Department and his subsequent moves to the Management Department and the West End Office Agency Department have lead him to chose the following optional competencies:

- Development appraisals.
- Purchase, disposal and leasing.
- Real estate management.
- Asset and investment management.

Mandatory competencies

James has progressed well with his mandatory competencies. The past 12 months of experience have ensured that his day-to-day tasks have allowed him to build on his strong existing interpersonal skills along with developing his understanding of the business and professional practice. These are clearly competencies that can be continued to be built on and improved over the following 12 months, but importantly he will enter into his second professional year with a broad breadth of experience and a strong understanding of the work that he has been involved in to date.

Core competencies

Valuation – James's valuation experience has been focused within his period in the Investment Department along with the West End Office Agency Department. His training has covered many aspects of valuation, including valuation for secured lending, valuation for investors, valuation for owner-occupiers, portfolio valuation and valuation for plant and machinery. His day-to-day work has improved his understanding of basic valuation principles, including rental analysis, yield selection and inspection and measurement. Importantly James has shown great aptitude in understanding how the other competencies come together when carrying out a detailed valuation on behalf of a client. It is our intention that during James's second year it is important for him to gain more experience in this particular competency, including more detailed Red Book valuation work. I am satisfied that James has achieved level 2 in this competency.

Inspection – James's role in each of the three departments that he has been in has involved the inspection of properties from different areas of the commercial property market. Each valuation he has undertaken has required an inspection, and this has allowed James to gain experience in measurement, identifying relevant building details and note taking. His role within the Office Agency Department has exposed him to analysing comparable properties from a property marketing or acquisition prospective. I am satisfied that James has achieved level 2 in this competency.

Measurement – Within James's experience in the Investment and Office Agency Teams, he has been exposed to a large volume of measurement work, particularly due to the fact that he was involved in the acquisition of a 350,000 sq ft investment building. James has also been involved in a number of property measurements within the West End Office Agency Team, on both acquisition and disposals and has had experience in relating these measurements to the *RICS Code of Measuring Practice* (5th ed). I am satisfied that James has achieved a level 2 in this competency.

Optional competencies

Development appraisals – James has been involved in a limited number of development appraisals, including outline appraisals on two buildings where a potential change of use was investigated. Of the two projects he has participated in, one was a comprehensive redevelopment (new build) and the second was a major refurbishment within the existing structure of the building. I am satisfied that James has achieved level 1 in this competency. However, it is our intention to expose James to more development appraisal work during the course of the next 12 months.

Purchase, disposal and leasing – James has been actively involved within the West End Office Agency Team, and has grasped a fundamental understanding of the market mechanisms at an early stage. James has been involved in the advice and recommendations, particularly in respect of marketing, to clients on the disposal of buildings. James has participated in written reports and also direct client meetings in respect of marketing and advice for both disposal and acquisition of office property. James has been involved in his own smaller projects, undertaking reports, site inspections and negotiations on behalf of office tenants in the acquisition of new premises for them.

I am satisfied that James has achieved a level 2 in this competency.

Real estate management – During James's period with both the Investment and Management Teams he has been exposed to a number of real estate management issues. It has been particularly important for him to work within both these teams as it has exposed him to the importance of communication and understanding between the two disciplines. During James's time within the Management Department, he has been exposed to a broad experience of all management issues, including dealing with tenants, dealing with contractors, undertaking service charge budgets, collection of rent and service charges and chasing arrears and debt collection. He has also been involved in liaising with tenants in respect of obtaining landlord's consent for assignments, subletting and tenant's fit-out works, including documentation such as Licences to Alter. I am satisfied that James has achieved a level 3 within this competency.

Asset and investment management – During James's time with both the Investment and Management Team he was actively involved in asset and investment management particularly focused on the Berkeley Square Estate. The Estate in particular has involved James in negotiations to take a surrender of an existing tenant's lease, and then a regrant on part of the original accommodation to a new tenant (associated with the former tenant) and achieving higher rental income, together with providing an opportunity for reletting the vacant accommodation. As part of the asset management experience, he has worked on the transition of securing vacant possession of ground floor units within part of the Berkeley Square Estate in order to change these from office to retail use, thus enhancing both the rental value and capital value of the assets on behalf of the client. This has involved liaising with existing tenants, clients, client's solicitors and new tenants and their advisers. I am satisfied that James has achieved a level 2 competency in this discipline.

Counsellor's comments

James can look back on his first year with some considerable satisfaction. He has gained some initial experience in the investment department, spent some six months in the management department and is working very successfully now with our West End Office Agency team. I expect him to complete his time in the investment department next year.

The wide breadth of experience that he has obtained to date has been under varying levels of supervision. At all times I have been satisfied with his performance, and his value to the firm has grown over the year as his experience and professional knowledge has expanded.

James has always been keen to develop his understanding of the profession, and his keenness to learn has been well received.

Candidate's comments

I am extremely pleased with the progress that I have made over the past 12 months and feel that my first year of practical experience has been extremely beneficial to me, and I hope also to my employer. I have made significant progress in order to achieve my goals with regards to the APC and the required levels at each competency.

Within the 12 months I have had a broad range of experience, spending a small amount of time in the Investment Department, a larger amount of time in the Management Department and then a short spell in the West End Leasing Department. Although so many changes of departments is not the norm, I feel it has given me the opportunity to gain a very beneficial insight into many different aspects of the property industry. I have also been fortunate enough to work within small teams as well as with large organisations and seen the disadvantages and advantages of both. I am looking forward to the next year, and spending more time within the West End Leasing Department as well as returning to the Investment Department.

I would like to take this opportunity to thank both my supervisor and counsellor for all their input and time spent on making my first year as easy as possible; at any point guidance was always on offer and everyone within the whole team was always happy to provide any assistance they could.

Chapter 5

Final Assessment Submissions

Final assessment submissions provide an opportunity for candidates to create a good initial impression on assessors.

These will usually comprise:

- Critical analysis – a 3,000 word report on a case in which the candidate has been actively involved.
- Summary of progress – a 2,000 word report summarising the experience gained since interim assessment (and accompanying supervisor's and counsellor's report).
- Log book since interim assessment.
- Professional development record since interim assessment.
- Record of progress.
- Interim assessment records.
- Application and other forms.

The diary also needs to be up to date, and the three-monthly supervisor's and six-monthly counsellor's reports in place, although these documents are not submitted to RICS.

There may be differences for some candidates, depending on their circumstances. These include whether candidates submitted interim reports to RICS under a previous APC system, whether candidates have previously been referred, and whether routes such as experienced, expert and academic/research contain differences. As always, candidates need to check the precise requirements.

Prior to the submission of the above documentation, candidates need to apply to RICS in order to take final assessment. Current application deadlines for the Commercial Property route are 15 December for the Spring interviews, and 15 June for the Autumn interviews. Deadlines to subsequently send the final assessment submissions are 15 January and 15 July respectively. Dates vary slightly for the Planning and Development route.

RICS send out the application forms automatically, although if they have not been received four weeks prior to the application date, candidates should contact RICS. The forms should however be available from the RICS website (as are templates for the summary of progress, log book, professional development record, etc.).

Essence of final assessment

As mentioned in Chapter 1, APC Overview, the final assessment stage of the APC is geared towards the experience being gained by candidates. Although the APC can appear complex on an initial examination of the RICS guides, the structure is necessary in order that every candidate can be assessed having regard to the nature of their employer's business and to the day-to-day case work in which they are involved.

There is still, however, a certain breadth and depth of knowledge required to be a chartered surveyor, which may not be gained as part of the case work available. All candidates have gaps in experience to some degree, and some seek to redress this more pro-actively than others. The importance of candidates viewing the APC as a training period of two years cannot be stressed enough; regular learning/ professional development activity being the key to first time APC success.

Being ready for final assessment

One of the important messages contained in *How to Pass the APC* is that candidates should submit for the final assessment interview only when they have gained the necessary experience, and reached the required competency levels. They must also be satisfied that there is sufficient time to prepare for the interview, having regard to any work or domestic pressures.

Too many candidates sit the interview when they are clearly not ready. Candidates should be proud to pass the APC first time, and should not work to the sometimes cited approach of, 'even if I fail, it will be good experience for next time'. In the eyes of colleagues and managers, APC failure can raise questions as to whether the candidate is really of the calibre they thought. It is rare that candidates lose their jobs as a result of failing the APC, but for some firms, the APC may work closely alongside performance management functions.

Whether or not APC failure leads to problems with an employer depends on the reasons for referral. If a candidate is unsuccessful, the chairman of the panel, in line with the views taken from the two other assessors at the end of the candidate's interview, will prepare a report outlining the reasons why. If a candidate has undertaken numerous lettings and sales as part of their experience, but is unaware of the agency law which underpins such work, or is unfamiliar with the *RICS Estate Agency Manual*, it would appear that a lack of effort and application is the reason. If, however, assessors are not satisfied as to the candidate's working knowledge of rent review processes, and the candidate has only been provided with a limited range of landlord and tenant work, the employer is likely to be more supportive.

One of the key aspects for candidates in determining whether they are ready to take the final assessment interview is 'knowing what they need to know' having regard to their personal circumstances (i.e. APC route, competencies, critical analysis, etc.). Unfortunately, for some candidates, the actual APC interview can be an insight into a range of issues they have not picked up through work experience or interview preparation.

Candidates who do fail are given a fair opportunity to start from scratch at their next interview, (even though a previous referral report can give a poor initial impression to assessors). Some referral reports, are, in fact, very encouraging, such as for a candidate who was judged by assessors to be hard working and conscientious, but falling slightly short of the required standard.

Another factor for candidates to consider is that if they demonstrate particularly poor application and knowledge at their interview, the assessors could require a further 12 months of experience to be gained before re-sitting, whereas a re-sit would usually be at the next six-monthly round of interviews.

Selecting a specialist area

Within the application forms for final assessment, candidates state factors such as their areas of experience, competencies and property type commonly dealt with (residential, commercial, etc.).

A particularly important aspect requiring consideration is the 'specialist area'. Options will be listed on the forms, and for the Commercial Property route the main ones are currently Valuation, Estate management/Landlord and tenant, Taxation and statutory valuations, Property investment, Property marketing and acquisition, Development and planning and Telecommunications. The main areas for the Planning and Development route are Planning, Development appraisals and Valuation.

Candidates' specialist area, among other factors, helps RICS ensure that candidates' experience is suitably reflected on the day of the interview by selecting assessors with the same areas of experience. This is achieved by assessors similarly advising RICS of their specialist area.

A consequence of the specialist area selected by candidates is that assessors' expectations of candidates will be greater in their specialist area, than perhaps in a third or fourth competency in which extensive experience has not been gained.

For example, a candidate who has undertaken a large amount of Red Book valuations, mainly for loan security and company accounts purposes of investment properties, may select the specialist area of Valuation. Particularly detailed questions may be asked in respect of the elements of the Red Book which related to the case work being undertaken, and regarding the use of quarterly in advance valuation techniques, investment market trends, yield selection, and evaluation of covenant strength through the analysis of company accounts and use of data from specialist credit rating agencies.

While the candidate who specialises in Estate management/Landlord and tenant would need to be familiar with the purpose and key content of the Red Book, and the fundamentals of investment valuation methodology, the interview questions in respect of valuation will be less detailed than the those received by the valuation specialist.

The estate management/landlord and tenant specialist would, however, be expected to have a more detailed understanding of rental valuation. This is because their rent review and lease renewal work would involve the precise pricing of individual factors affecting value; examples including the effect of a restrictive user clause, a long unexpired term, or flexible break clause facilities for the tenant.

Summary of progress

The summary of progress at final assessment is a 2,000 word report outlining the experience gained since interim assessment. This is similar to the summary of progress undertaken at the interim stage, and involves experience being set out under competency headings (plus an introduction, if required, and a heading of professional development).

The summary of progress is completed on the RICS templates. There may, however, still be candidates working to an older APC system who are required to undertake a 1,500 word summary of experience in a traditional report format, instead of the template versions.

It is important that the summary of progress at final assessment picks up on points mentioned in the interim assessment reports – in particular the forward plan which outlined how the necessary experience was to be gained in the period following interim assessment.

As mentioned in Chapter 4 in respect of interim assessment, it is important that candidates convey the due depth of their work to assessors as well as the breadth of experience gained. This can be achieved by reference to case examples, including, for example, key negotiating points or factors affecting value, and legislation and case law which may have been drawn on.

As with the work experience outlined in the interim assessment reports (and indeed with all aspects of final assessment submissions), candidates must remember that some interview questions will be based on the information that has been submitted.

Where candidates still have to gain experience in certain areas prior to sitting the interview, they can make suitable entries in the 'training planned' section of the RICS summary of progress templates.

Having provided an example of a summary of progress at interim assessment in Chapter 4, and candidates having already been through a similar process by the time they reach final assessment, a further example is not included here.

Competencies, log book, etc.

By the time of final assessment, candidates should be well-organised in respect of diary recording, log book completion, the professional development record and the record of progress.

If further guidance is needed, reference can be made to Chapter 1, APC Overview, and Chapter 2, Starting the APC. It is, in fact, worthwhile for all candidates to review the APC requirements, and to be satisfied that these are being met as part of final assessment written submissions.

There are, for example, occasional candidates who do not obtain their supervisor's and counsellor's certification that they have reached level 3 in a particular competency, and are questioned by the chairman on the day of the interview about such discrepancies. While the RICS staff will check matters such as candidates' eligibility to take final assessment, they will not scrutinise submissions to ensure that candidates have selected the correct competencies, been signed off by the supervisor and counsellor etc. Assessors will receive candidates' submissions around four weeks before the interview, and if discovering discrepancies may wait until the interview. The assessor's approach does, however, depend on the extent of any discrepancies. It is rare that an interview does not go ahead, but some discrepancies could contribute to a referral.

As candidates' submissions to RICS will be some time prior to their interview, post-submission experience continues to be recorded in the log book, which is taken to the interview. For some candidates, this will be an important means of demonstrating that the necessary experience has been gained. This could be the case, for example, with a candidate working for one of the larger practices who rotate candidates around departments, and at the point of submission, the candidates has two to three months of experience to gain in a particular department. The professional development record similarly continues to be recorded, and is taken to the interview.

In some cases, supervisors and counsellors may feel unable to sign a candidate off to a required level at the point of submission to RICS, but consider that the two to three months of further experience (and surrounding learning) prior to the interview, will see the candidate reach the necessary level. Candidates can therefore state in their summary of experience that this is the case, and it is also helpful to add a comment to the record of progress to this effect (otherwise it could appear uncompleted). On the day of the interview, a letter can be taken, confirming that the necessary levels have been met, as certified by the supervisor's and counsellor's signatures.

For the interview, it may also be helpful for candidates to take their critical analysis and any other records. Occasionally in the interview, this may provide a helpful reference point to deal with a difficult question.

Report writing and presentation

As with all the submissions, the required standard of report writing and overall presentation needs to be demonstrated to assessors. Brief guidance is provided in Chapter 12, and any candidates requiring further support in this area can refer to a range of business and management literature available through high street bookshops.

It is important that candidates leave sufficient time to prepare their written submissions in order that they are of the best possible quality. Hurried submissions, where candidates have not allowed sufficient time for reflection and restructuring, can come across to assessors as poor quality first drafts. Some firms require candidates to submit their work to supervisors and counsellors one month in advance of the RICS deadline. This ensures that supervisors and counsellors have time to properly consider candidates' work and provide the necessary support.

Preparation for the interview

As well as actually being ready for final assessment in terms of the experience and knowledge gained over the training period, candidates must be familiar with the interview assessment process they are about to face, and prepare accordingly.

Chapter 8 provides an overview of the APC interview, and Chapter 9 provides guidance on the presentation delivered on the day. Chapter 11 includes examples of interview questions and answers. Reference should also be made to Chapter 10 which provides illustrations of aspects of practice which need to be considered in relation to candidates' competencies.

Chapter 6

Critical Analysis

This chapter has been prepared in conjunction with Advantage West Midlands, the regional development agency for the West Midlands. Advantage West Midlands has recently facilitated free of charge APC events for general practice and planning and development candidates as part of their 'Advantage West Midlands CPD Centre' initiative.

The co-author of this chapter, Kitt Walker of Advantage West Midlands, is a former APC candidate.

Effect of the critical analysis on the interview

The critical analysis is the APC candidate's principal opportunity to demonstrate experience and capability to assessors in advance of their final assessment interview. On the day of the interview, it forms a 10 minute presentation by the candidate to the three assessors, and is the subject of around 10–15 minutes' detailed interview questioning.

The critical analysis is a report on a case in which candidates have been actively involved (although RICS refer to a 'case' as a 'project'). It comprises a maximum of 3,000 words, plus appendices.

Factors affecting the choice of critical analysis are set out below, together with guidance on its content. Chapter 7 provides an example of an APC candidate's critical analysis.

Familiarity with the subject

Candidates should have gained detailed experience in the subject of the critical analysis, and not use a case relating to a subject or property type in which they have had limited experience: possibly confined to the critical analysis itself.

It would not, for example, be wise for candidate to undertake a critical analysis on a rating appeal in respect of a bank, if the rating competency (Local Taxation/Assessment) has been taken to only level 2, and/or limited experience has been gained regarding retail properties, including banks. As with all areas of practice and property types, there is knowledge and understanding developed to a certain level only once actively involved in such sub-markets. Had the candidate been involved in industrial agency and also having gained experience in undertaking investment valuations in the industrial sector, a critical analysis could be undertaken on an industrial letting, or a Red Book investment valuation of an industrial estate, for example.

A critical analysis will often be in candidates' stated specialist area, such as Valuation or Estate management/Landlord and tenant, but it does not have to be.

Importance of active personal involvement

It is important that candidates have had an active personal involvement in the case, and are not tempted to submit a project which they hope will impress assessors through being high profile and/or of high value.

To be able to account for the decisions undertaken within a case, and thoroughly understand all elements, candidates will need to have been involved actively, as opposed to providing a helping hand to colleagues. Also, the critical analysis is the candidate's opportunity to demonstrate that they are at the necessary level to be chartered surveyors. This will not be achieved by commenting at undue length about the case background, other colleagues' work, etc.

The overuse of 'we' and 'our' in candidates' submissions is an example of how candidates can easily give away their lack of personal involvement to assessors. Assessors can also gauge the true level of responsibility that candidates would be given in dealing with clients, or acting on behalf of an employer on an in-house basis. Within the critical analysis, candidates need to be clear about the aspects for which they were personally responsible, and any elements which required particular input from colleagues.

Simplicity and complexity

The often heard advice of 'just do something simple' is partially correct, but as candidates are looking to demonstrate that they have reached the level of chartered surveyor, cases which are too basic will not demonstrate this, and may also raise questions as to the overall quality of work in which candidates are involved.

The critical analysis really needs to be a balance of not being too simple and not being too complicated. A case which is too complicated is one where too much descriptive information is needed in order to suitably account for decisions which were undertaken.

The critical analysis need not cover all of the candidate's core and optional competencies, although it is bound to relate to some. In fact, a critical analysis which attempts to cover a wider range of issues will usually fail to convey a sufficient level of depth in any particular area. This is why it is important for candidates to be selective about the areas on which they concentrate in the critical analysis, and not take the alternative approach of feeling they have to account for every aspect, however small, which arose during the case.

One or more cases

Although RICS states that one or more cases could be chosen, only one case tends to be necessary, unless the essence of the critical analysis is the link between cases. Examples include an overall estate management strategy for a portfolio covering several buildings, or a rent review strategy covering a whole parade of shops because of the interrelated review, investment and possible letting issues for each unit.

The covering of two separate projects could, of course, raise questions as to whether sufficient experience has been gained to be able to cover an individual case.

Timing, and case completion

The case has usually been undertaken since interim assessment. Candidates' knowledge has by then developed sufficiently to explore the issues in detail, which is unlikely to be the case in the early stages of the training period.

The case has ideally been completed in order to provide an outcome upon which analysis can be based. However, if a certain stage has been reached with an incomplete case, this may still be suitable for a critical analysis – first, because there will still be an outcome/achievement to facilitate analysis, and, second, as it is almost always necessary to focus on elements of cases in order to provide the right depth of comment within a limited word count.

If, for example, a marketing case was, 'The marketing strategy formulated for the sale of an office investment', a good quality critical analysis may be confined to bringing the property to the market, covering method of disposal, choice of advertising outlets, observing agency law, lease analysis, evaluation of tenant covenant strength, and valuation. The level of initial interest received, and a brief overview of early discussions with applicants would enable a reflective analysis (see later) to be provided, and it would not be necessary for the case to reach an exchange of contracts. Planning and development cases, as well as construction cases for building surveyors and quantity surveyors, are a good example of where a case within a wider instruction needs to be carved out for the critical analysis.

Where cases have not been completed, this may raise issues of confidentiality which make the case inappropriate for the critical analysis. Candidates should always establish from their supervisor and counsellor, at an early stage, whether the entire case, or parts of the case, are too confidential to cover. Some information can be sensibly withheld, such as a client's name, but the absence of a location plan, photograph and comment on the client's objectives, would usually take too much away from the critical analysis.

Where an incomplete case still contains sufficient detail to be suitable for the critical analysis, it can contain a brief summary of outstanding points, and how they will be approached. The case will, hopefully, be complete by the time of the interview, and the presentation can provide an appropriate update, including any revised evaluation of the work undertaken.

Critical analysis, not narrative

It is always important, especially when writing the critical analysis, that candidates remember that it is a critical analysis, and not a more passive narrative of events. This involves critically analysing the facets of the case being reported. This will be different from the type of reports that candidates may be preparing as part of their day-to-day work.

Some candidates provide only an overview of events, and simply report what happened. Here, there can be extensive descriptive commentary on the property itself, undue general commentary, such as on the importance of valuations being accurate, and inappropriate lecture type commentary such as a summary of the UK's environmental and contamination related legislation. It is imperative that the critical analysis is case specific, and its contents best sell a candidate's experience and abilities to assessors. As indicated below, important aspects include the identification of the client's objectives, highlighting key issues, evaluating options and providing a critical appraisal/reflective analysis of the work undertaken.

On the RICS referral reports, outlining the reasons why candidates have failed, assessors sometimes state, 'the critical analysis was not a critical analysis', and add comments on the lines of the above regarding the unduly descriptive nature.

RICS requirements: use of headings

As always, candidates need to scrutinise the RICS APC guides as part of the preparation of written submissions.

For the critical analysis, the edition 1/July 2002 guides (current at the time of publication – as always noting the possibility of future changes) include comments such as those above, and also state that candidates need to include the following headings:

- Key issues.
- Options.
- Reasons for rejection of certain options.
- Your proposed solution to the problem(s) and reasons for supporting this choice.
- A critical appraisal of the outcome and reflective analysis.

This helps steer candidates towards producing a critical analysis rather than a narrative. It is important that candidates do not structure their report too rigidly in relation to these headings, such as by simply adding a heading of 'Introduction' to the above list, and then trying to fit their account of their work into the headings. This can prevent their report running chronologically in line with events as they occurred. It is worth noting that a chronological account of events tends to be most captivating for the reader/assessor.

The required headings will be drawn on by each candidate in a different way, depending on the nature of their case. One candidate may see a whole case centre around a major option – such as the decision to obtain outline planning permission before marketing a residential development site, or, alternatively, seeking expressions of interest from developers, with a view to granting an option agreement for the developer to thereafter secure planning consent themselves. Another candidate may have a smaller series of options, such as in a rent review where there may be tactical issues and options of whether to serve a rent notice immediately, or wait for improved rental evidence to emerge, whether to serve a calderbank letter, or whether to opt for arbitrator or independent expert, if the lease permits such a choice.

Expressing the reasons for taking a particular course of action is an important means of candidates conveying to assessors that they fully understand the work in which they have been involved – as opposed to, for example, just responding to a superior's instructions without much thought. It would not, for example, be appropriate to only state that it was decided to sell the property by auction, and not acknowledge the alternatives of private treaty, informal tender and formal tender. The need to account for actions in such a way, rather than only report on events, adds to the word count. This is also why candidates may initially consider that it will be a struggle to find 3,000 words, but after evaluating options and meeting other analytical criteria of a good critical analysis, can see that they can reach 5,000 or more words on a first draft. Sometimes this leads to the refining of the scope of the critical analysis, and also its re-titling. The disposal of a residential development site, for example, may become the marketing strategy implemented for the sale of residential development land. This would allow the critical analysis to focus on the marketing aspects, and give less attention to planning, development appraisal and negotiations, for example.

The critical appraisal of the outcome and the reflective analysis of the experience gained are important elements of the report. The critical appraisal can comment on whether the outcome successfully met the client's objectives, which should be outlined at the start of the report. Comment can be made on any aspects which could have been done differently. While the report is called a critical analysis, and a section of critical appraisal is required, good candidates often find little scope to be critical. Nevertheless, suitably articulate comments can still be made which vindicate the approach taken. Candidates should not be too critical, and cases involving mistakes, albeit possibly a good learning exercise, should be avoided. Reflective analysis is a more personal commentary on the experience gained, including learning outcomes. This should be confined to precise points, and not, for example, a commentary in relation to each core and optional competency.

Overall headings are the choice of the candidate, subject to meeting the above stated RICS requirements in respect of headings. This is, however, an area where, in practice, assessors are not unduly concerned about the precise headings adopted, provided the report meets other requirements, especially candidates' demonstration of their proficiency in practice – including by way of a genuine 'critical analysis' rather than narrative.

Other requirements from the RICS guides which are current at the time of publication comprise the inclusion of photographs and plans, the signing and dating of the report by the candidate, and also certification by the supervisor and counsellor, as well as stating the word count (which can include comments such as whether headings have been included or excluded).

Prompting interview questions

In preparing the critical analysis, an eye needs to be maintained on the position that candidates could be in under interview questioning. Statements such as, 'I ensured that the works were undertaken in accordance with all relevant health and safety legislation', or 'after checking the relevant case law, I maintained my previous position' may well see assessors asking about the relevant legislation and case law.

Some candidates deliberately leave in such 'hooks' in the hope that assessors are led into questions that can be met with strong pre-prepared answers. Assessors are aware of this, and where hooks are blatant, tougher questioning could be focused on other elements of the submission for which candidates may not be so well-prepared. Tactical gaps could also be regarded as an absence of essential information – as well as the poor impression created by trying to hoodwink assessors.

Key interview questions

Candidates can sometimes be caught out with questions such as:

- What was the rateable value of the property?
- What were service charge rates per sq ft?
- Was VAT added to the sale price/rent and why?
- Did you investigate any alternative use potential?
- What use class was the property in?

These can raise questions as to how actively or thoroughly a candidate has been involved in the case.

Confidentiality statement

A statement on the lines of 'The consent to disclose the information contained within this report has been obtained from my employer and the client' (assuming it has), demonstrates professional awareness. But statements such as 'It is requested that the panel do not divulge this information to third parties', are inappropriate, as APC submissions are clearly confidential to the panel, and there should not be any suggestion that they may not be treated as such.

Adherence to word count limits

When RICS states that 3,000 words are required, compliance is expected, and while assessors may not be concerned about a stated word count slightly above, non-compliance raises questions about a candidate's ability to similarly follow client requirements – as can non-compliance with any elements of final submission.

Making best use of appendices

The appendices should suitably complement the information provided in the main report, rather than, for example, provide further commentary because of the word count limits for the main report.

Typical examples of appendices for a critical analysis prepared by a general practice surveyor include:

- Location plan – such as an Ordnance Survey 1:1250 or 1:2500 plan showing surrounding features.
- Photographs.
- Plans (if relevant, such as showing the internal layout for zoning, but not all floor plans simply for being comprehensive if there are no specific case issues to which they relate).
- Lease extracts – but not a full copy of the lease.

- Schedule of comparable evidence – listing details of each transaction, with an overall commentary being provided in the main body report.
- A valuation – unless more easily included in the main report where sufficiently brief (such as area × rate per sq ft/m² for rental valuation, or YP perp. for capital valuation).
- Copy of a planning brief/development brief.
- Extracts from the RICS Red Book, Code of Measuring Practice etc. – but, again, which are relevant to precise points raised in the report.
- Extracts from case law or legislation – again, which are relevant to precise points raised in the report.

Report writing skills

As well as creating a poor initial impression on assessors, inadequate report writing skills are grounds for referral – as chartered surveyors must of course be able to communicate effectively in writing, as well as verbally. Spelling and grammatical errors are obvious downfalls, as it the common failure to include apostrophes, such as in 'the client's objectives'. '1990s' does not have an apostrophe, neither does 'dilapidations'.

Poor written skills also involve using too many words when considerably fewer would say the same thing, and the poor structuring of a report. The fewer the number of words used in a report, the more contented the reader with the expediency and impact of a report. In the case of the critical analysis, where the word count is limited, the tighter the text, and the greater the scope to demonstrate the depth of knowledge sought by assessors.

As an example of the unduly wordy text found in a critical analysis, a candidate may write:

> My first task was to undertake a check to ensure that there were no conflicts of interest in relation to the client or the property. It was established that there were no conflicts.
>
> It was then necessary, as with all such cases, to prepare a letter of instruction detailing the terms upon which we would act for the client, and to ensure that the client was fully aware of the service that we were providing. The client confirmed acceptance in writing within one week, and it was then possible to proceed with the instruction.
>
> My first task was then to inspect the property. ...

Faults include unnecessary information, the impression that the detail is a big deal to the candidate (whereas much is taken as read at this level), half-repetition of previous sentences, similar words unnecessarily used (such as 'undertake a check to ensure that') and finer points such as 'first task' used twice. A more polished version could be:

> After checking that there were no conflicts, and agreeing terms of instruction, I inspected the property. ...

This has taken 16 words instead of 103 – although more detail than these 16 words would still be acceptable. Interview questions could include, 'How exactly did you check there was no conflict of interest?' or 'Give an example of what a conflict could have been?' Assessors may also ask about individual terms of instruction.

Also, it is not necessary to begin the critical analysis on the lines of the report being produced in accordance with a particular paragraph and edition of the RICS guides. The reader/assessors can be better endeared to candidates' work by sharp, fluent text. It also looks poor, for example, when the edition referred to is several years out of date, and not the correct version for the candidate to be following.

'Write for the reader' is a big message for candidates, with many writing for themselves, and missing out or misplacing information which leaves the reader frustratingly turning back, or turning forward, pages in order to piece things together – and sometimes never finding essential detail.

Report writing and presentation standards are also covered briefly in Chapter 12.

Re-writing existing material

Another important element of report writing, as well as being an issue of professionalism and professional ethics, is that the report should be written by the candidate, and not comprise the copying-and-pasting of written work which may have been prepared as part of the case, or separately.

Property descriptions and market commentaries are the biggest giveaway, such as comprising estate agent speak from marketing brochures, or incisive economic commentary from research publications (and which sometimes has no relevance to the case).

Assessors can easily spot a change of written style which derives from copy-and-pasting from other pieces of work, and will judge candidates accordingly as part of the decision whether they should pass or be referred.

Chapter 7

Example of Critical Analysis

This section provides an example of a candidate's critical analysis. As well as highlighting aspects relating to APC written submissions requirements, the report provides a useful insight into the work that may be undertaken in practice in a particular subject.

The critical analysis is prepared by Paul Richardson, who passed the APC in Spring 2003. Title and contents pages, signatures and some appendices are excluded, and some numbers are altered. Headings have also been altered, and the example can be used as an exercise by other candidates in considering whether RICS requirements in respect of headings and content have been met, and, if not, how any adaptations can be made. The text has been condensed, but as an illustration of a suitable form of presentation, 12 point Times New Roman style, with 1.5 line spacing, was adopted. Client details and location factors are altered.

1. Introduction

The subject of this critical analysis is a rent review of an industrial unit in Hallsworth. This is one of the instructions I have completed this year acting for the Tenant, and I have selected this instruction for several reasons:

- The instruction falls within my area of expertise.
- The review reflects the type of work I have undertaken this year.
- The level of responsibility and involvement which I had were high and, therefore, allowed me to demonstrate and develop a wide range of skills.
- The scope of the instruction was wide, covering to varying extents all my competencies as well as broader surveying issues.

The instruction

The instruction was received from our clients, Cavanagh Building Supplies, following the service of a trigger notice from the landlord. The objectives of the instruction were to:

- Ensure that the trigger notice was valid and that the rent was being reviewed on the basis prescribed in the lease.
- Negotiate the best possible rent on behalf of the tenant.
- Avoid causing any prejudice to any subsequent rent reviews.

Achieving these objectives involved the following procedural steps:

- Analysing the lease.
- Inspecting and measuring the property.

- Researching comparable rental evidence and local/regional market trends.
- Valuing the property.
- Reporting and advising to the client.
- Negotiation with the landlord.

Description of the property

The property is located in an established industrial location on the outskirts of Hallsworth approximately 4.5 miles from the motorway junction (Appendix I, II).

Park Close Industrial Estate on which the property is situated comprises 14 units of identical age and construction, offering B1/B8 accommodation between 131.45 m² and 403.46 m² (1,415 sq ft and 4,343 sq ft).

The property comprises a single storey terrace unit built in the mid 1980s. The construction is of steel frame with clad brick/block work walls beneath insulated steel profile upper elevations (see Appendix III).

Client confidentiality

I have obtained written permission from Cavanagh Building Supplies to use this case as the subject for my critical analysis.

2. Initial procedure

Client instruction

The instruction was received in writing via the principal surveyor within my employer who co-ordinates the national instruction to act on behalf of Cavanagh Building Supplies in all Landlord and Tenant property matters.

The basis of the instruction was confirmed in accordance with the existing terms and conditions of engagement contracted between my employer and Cavanagh Building Supplies under the national instruction.

In accordance with my employer's ISO9001 QA certification, a file was opened to hold all legal documentation and correspondence pertaining to the instruction. As requested, the client forwarded a copy of the lease documentation, the landlord's trigger notice and the client's counter notice.

Conflict of interest report

In accordance with the Estate Agents Act 1979 and the RICS Rules of Conduct, a conflict of interest report was conducted using my employer's instruction database. The report revealed no conflict of interest.

3. Analysis of the lease

Confirmation of documentation

Prior to conducting a detailed analysis of the lease I contacted the client. The client confirmed that the document I had received was the current lease under which the property was occupied, and that there were no other side agreements, or licences which had been granted in addition to the lease.

Validity of the trigger notice

I referred to the lease to ensure that the rent review trigger notice served by the landlord was valid. The main considerations were:

- Does the landlord have to serve a trigger notice to initiate the rent review?
- Are there any time constraints within which notice must be served?
- If there are time constraints, does the lease specify that time is of the essence, or are there any deeming provisions in the lease that would make time of the essence?
- Has the notice been served in the correct format containing the correct information?
- Has the notice been served to the correct address?

Following these considerations I concluded that the trigger notice was valid as time was not of the essence and there was no prescribed form of notice.

The effect of the counter notice

The client acknowledged the trigger notice through a 'without prejudice' counter notice, and advised the landlord that my employer would be acting on their behalf in this matter. The effect of this notice was to acknowledge receipt of the trigger notice with out prejudice to the tenant's right to challenge the validity of the notice.

Rent review clause

I referred to the rent review clause within the lease to identify the basis on which the rent was to be reviewed (Appendix V). This was a traditional rent review clause and, therefore, was not onerous in this market.

Third party determination

The lease provided that if the matter could not be resolved within three months of the service of the landlord's trigger notice, the parties may refer the matter to a third party acting as an Independent Expert for determination. This information was to be applied when considering my strategy for negotiating with the landlord.

Lease terms affecting value

As it must be assumed that the property is held on the existing lease I identified the terms within the lease that impact significantly of the rental value of the property (Appendix VI).

4. Property inspection

The property

The property was inspected and measured in accordance with the *RICS Code of Measuring Practice* (5th ed). The inspection identified the physical factors of the property that would effect the rental value.

Comparable property and the area

An external inspection was made of the other units located on Park Close Industrial Estate, and the occupiers were noted. A wide inspection was then undertaken of the surrounding area to identify the other types of property and occupiers located in the vicinity and to determine the quality of the infrastructure serving the property.

Floor area

Using my inspection notes and measurements taken on site I prepared a plan of the property and calculated the floor area. The floor area was calculated on a gross internal basis in accordance with the *Code of Measuring Practice* (5th ed).

5. Valuation

Comparable evidence

I researched details of rental evidence relating to comparable industrial property within Hallsworth. Research was undertaken through verbal and written enquiries to surrounding occupiers and commercial property agents. Details of the comparable evidence are shown in Appendix IV.

Hierarchy of comparable evidence

Rental evidence was derived through various market transactions. By referring to the hierarchy of rental evidence (Appendix VII) I weighted the evidence according to the strength of the transaction. By identifying the weaker sources of evidence at an early stage, I had time to undertake further research to get 'behind' the transaction. In this further research stage I considered the following:

- Was the tenant professionally represented?
- Did the tenant comply with all time constraints within the transaction to establish an even negotiating platform?
- Were any incentives offered?
- Is the tenant a special occupier?

Rental valuation

I conducted a valuation to Open Market Rental Value in accordance with the rent review clause within the lease.

Market conditions

My site inspection revealed that Park Close Industrial Estate was fully let and that none of the tenants were looking to assign their leases. A large mixed industrial/office development was under construction on brownfield land adjacent to Park Close Industrial Estate. Other industrial estates in the vicinity were also well let. These factors indicated a strong demand for industrial accommodation, which is being met by modern developments. This type of market with a high demand and limited comparable competition would suggest that the hypothetical tenant, to be assumed under the rent review clause, would be in a relatively weak negotiating position.

6. Client report, recommendations and strategy

I produced a report to the client describing the property, identifying the factors affecting rental value, providing an explicit rental valuation, and making recommendations for negotiating with the landlord.

Following receipt of my report the client reissued instructions to progress the review in accordance with the recommendations made in the report.

In view of the rental evidence it was clear that the current rent was reversionary and that the landlord would submit a strong case to justify a rental uplift.

Market research indicated that industrial rents were increasing. Furthermore, the adjacent industrial development was nearing completion with pre-lets available from £4.20 per sq ft on the larger 20,000 sq ft (1,858 m²) units. In view of these factors I advised that the client should act promptly in this matter, as any new rental evidence was likely to support a rental uplift.

The lease provided that either party may refer the matter at any time to a third party for determination of the rent. The third party would be required to act as an Independent Expert, and the cost of the Expert would be disbursed equally by both parties. In acting as an Expert rather than as an Arbitrator, the third party would be capable of analysing his own contemporary rental evidence and not reliant on historical evidence submitted by the parties. Accordingly, I decided that it would be in my client's interest to submit an early application. I recommended that a time scale for negotiations prior to an application should be agreed with the landlord as a matter of urgency.

I suggested liasing closely with the adjacent tenant who was in the process of renewing a lease to ensure that we would not prejudice one another's rental negotiations.

Following my valuation I recommended negotiating with the landlord with a view to settling at a rental in the region of £18,700 pa.

7. Negotiations

Landlord's view

The landlord appointed a chartered surveyor to act on their behalf in the rent review. The landlord's surveyor served correspondence indicating that the passing rent of £14,200 pa should be increased to £25,000 pa. Using my floor area calculations, this was an increase from £3.27 per sq ft to £5.75 per sq ft. The landlord's surveyor justified this increase with reference to the following:

- The unit comprised 5,050 sq ft (469 m^2) of accommodation (707.50 sq ft (66 m^2) greater than my floor area calculation).
- The comparable rental evidence relating to unit 1.
- As a trade counter user, the tenant would pay a higher rent for prominence to Hermitage Road.

Negotiation and counter argument

The principal areas of negotiation were as follows:

Floor area – Given the difference in our floor area calculations, I requested a copy of the landlord's surveyor's measured survey. The plans confirmed that the surveyor had included the mezzanine floor within the floor area calculations. Under the terms of the lease, as the tenant had installed the floor under a licence from the landlord, the works were to be disregarded at rent review. Thus, we agreed my floor area of 4,343 sq ft (403 m^2).

Comparable evidence – The landlord's surveyor based his valuation on the £4.92 per sq ft achieved on unit 1. My research revealed that the tenant at unit 1 was not professionally advised in agreeing the rent at review. The evidence did not fit with the Open Market letting of unit 10, which was achieved in the following month. I argued that under the hierarchy of evidence, the open market letting evidence should be given more weighting. I added that the tenant at unit 10 already occupied a unit on the estate, and may be regarded as a special purchaser that may pay a rental above market rent. Accordingly, the landlord's agent agreed to reduce the weighting on the evidence from unit 1.

Use and prominence – The landlord's surveyor argued that the use of the subject property as a trade counter would benefit more significantly from frontage to the busy Hermitage Road than the adjacent traditional industrial occupiers. I referred to the Use Classes Order 1987 to confirm that the property was being used in accordance with Use Class B8, as specified under the lease. Researching comparable evidence revealed that all units on Park Close had the benefit of B8 use. We agreed that a rental adjustment should be made to reflect the prominence of the unit to Hermitage Road, however, the allowance was to be no more than for the allowance for other units on the estate.

Quantum – In the proposal, the landlord's surveyor had not made any adjustment to reflect the size of the subject unit in comparison to the other units on the estate. I, therefore, made the case for a quantum allowance. The landlord accepted my argument and proposed to make a 2.5% rental allowance to reflect quantum. I referred to the comparable evidence and identified transactions where the size of the units was the only variable and calculated the ratio between the percentage difference in the size of the units and the percentage difference in the rent. By assuming a linear extrapolation, I applied the ratio to the subject unit. The landlord's agent agreed to a quantum allowance of 4.4% derived from my ratio analysis.

Lease Term – In my valuation, I included a rental allowance to reflect the 10-year lease term which was to be assumed under the hypothetical review clause within the lease. The rental evidence indicated that the market trend was for shorter leases between 5 and 6 years, often incorporating tenants' break options. I, therefore, took the view that a 10-year lease term was onerous on a tenant, and that a hypothetical landlord would accept a lower rental in consideration for a longer secure income stream. I supported this argument with rental evidence and confirmed my intention to refer this matter for third party determination.

Stepped Rent – During the negotiations the landlord's agent suggested agreeing a stepped rental over five years (i.e. two years below the agreed value, one year at the agreed value, and two years above the agreed value) to accommodate any rental uplift. I realised that by applying a discount rate to reflect the time value of money, the tenant would pay less over the five years. Despite this advantage, I chose to recommend to my clients that this offer should be rejected as the artificially high stepped rent at the end of the five years may prejudice subsequent rent reviews on the subject unit, and also on the other units located on Park Close.

Agreement

Through negotiation, the rent was agreed at £18,500 pa (Appendix VIII).

The lease provides that the review must be documented by the joint signature of a rent review memorandum to be produced by the landlord. I ensured that the memorandum was in the prescribed form and that the correct lease details, dates, parties names and rent were included.

The fee was submitted in accordance with the terms and conditions agreed upon the acceptance of the instruction.

8. Reflective analysis

Experience gained

My work in this instruction allowed me to develop and consolidate the following areas and skills:

- The process of receiving and servicing an instruction in a professional manner. This developed my communication skills and report writing skills.
- Identifying procedural steps within a rent review and following them under QA recommendations. Using the procedural steps enabled me to organise, prioritise and manage my time effectively to provide an efficient service.
- Analysing of the lease documents enabled me to become more familiar with the standard terms and layout of a lease.
- Understanding the client's business within the B8 user clause and be aware of landlord's potential interpretation of the use.
- Efficiently collecting and analysing comparable rental evidence, being aware of the various sources of rental evidence such as agents, occupiers, investment particulars, and web-based services.
- Understanding the process of negotiation and the importance of

preparation prior to negotiation. During actual negotiations it is important to identify agreeable points at an early stage, and to set out and work through the factors effecting value in an ordered manner.

- Considering reasons behind rental adjustments and how they may be defended or challenged through negotiation. I realised that where possible, it is beneficial to try and quantify through rental evidence any traditionally qualitative rental adjustments.
- The use of third party determination and the need to be proactive in this use. I realised how this may be used as a negotiating tool without actually referring the matter to a third party for determination.
- Confidence in my ability and knowledge to take my own decisions. I did however acknowledge the limit of my experience, and on a number of occasions referred to my supervisor to confirm my thoughts.

Areas for development

I met all my objectives efficiently within a self imposed time-limit and agreed a rental value lower than the value calculated in my report. Despite a successful completion, reflecting upon this instruction I have identified various areas for development.

Proactivity

Although the instruction was serviced in accordance with the terms of engagement agreed by my employer and Cavanagh Building Supplies, I believe that as a national client I could have serviced this instruction in a more proactive manner. Should the client issue the instruction to act in the review six months prior to the rent review date, I could have monitored the market and prevented any unrepresented tenant prejudicing my client's position with an uninformed rental agreement.

In being proactive in a case where time is not of the essence, it may be possible for the tenant to trigger the review – or alternatively make time of the essence in respect of the landlord having to trigger the rent review.

By being proactive in this instruction it could have been possible to identify the landlord's arguments for increasing the rental at an early stage. Improved anticipation of the landlord's arguments would have allowed me to identify immediate counter arguments.

Quality of rental evidence

I realised the need to collect as much detail in my rental evidence as possible. I found that I had to contact agents and occupiers more than once to collect additional information such as the user clause or the terms to be assumed under the rent review provisions.

With regard to the physical factors, I could have taken more photographs of the comparable units, in order to assist in my valuation and also for use in supporting rental adjustments.

Organisation

When researching comparable rental evidence, I need to improve my use of file notes and records of telephone conversations, in order to allow my colleagues to use the file confidently in my absence, and so that the file may be used for reference in future rent reviews.

9. Conclusion

I met the objectives identified at the beginning of the instruction and have gained experience of many aspects of rent reviews.

Identifying areas for development together with reflective analysis has allowed me to learn from my experience, with a view to applying the knowledge and experience gained from this instruction in future instructions.

As part of the wider 'picture', I now hold a better understanding of the role of Landlord and Tenant surveyors and their impact on a tenant's business or the landlord's investment interest.

I have developed my core competency area together with skills in other competency areas, and appreciate how the various competencies are linked.

I have developed skills to meet the objectives in accordance with QA procedures and the RICS Rules of Conduct in a professional manner.

Illustration of appendices

Appendix IV – Comparable rental evidence

After listing evidence under headings of Address, Tenant, Effective date, Type of transaction (e.g. letting, rent review), Rent per annum, Area, Rent per m²/ per sq ft, a remarks column included the following examples.

- Tenant was unrepresented. The assumed hypothetical term was five years. The user clause allows for B1 and B8 use.
- Areas not agreed – Renewal not complete. New lease for a term of six years subject to a tenant only break option in the third year. The user clause allows for B1 and B8 use.
- Unit located to the rear of the subject unit, thus does not benefit from frontage to Green Lane. The unit is held on a 15-year lease from September 1989. The user clause allows for B1 and B8 use.
- Design and build office and industrial units are available to let in an equally prominent location. The marketing agents are quoting rentals from £4.20 per sq ft.
- A quantum allowance must be made in considering this evidence.

Appendix V – Rent Review Terms

The basis of the rent review is:

- Five-yearly upward only rent reviews from the lease commencement date of 1 June 1987 (accordingly the review date is 1 June 2002).

- From the review date, the rent is to be reviewed to the greatest of the Rack-Rent payable immediately prior to the reversion date and the current Open Market Value.
- Open Market Value is to be determined assuming the property is available to let under the current terms of the lease, as a whole on the open market by a willing lessor to a willing lessee without fine or premium and for a term equal to the remainder of the term. The usual section 34 of the Landlord and Tenant Act 1954 disregards are provided in the lease.

Appendix VI – Lease Terms

The terms within the lease that impact significantly of the rental value of the property are:

- The term of the lease: The property is held on a 25-year lease from 1 June 1987.
- Repairing obligations: The tenant is responsible for both internal and external repairs.
- Decorations: The tenant must decorate the exterior and interior of the property in every third year.
- The User Clause: The permitted use is as a warehouse with ancillary offices within Use Class X (now B8) of the Schedule to Town and Country Planning (Use Classes) Order 1972.
- Alienation provisions: The tenant may only assign or sublet the whole of the property subject to receiving the landlord's prior written consent.
- Service charge: The service charge was clear of caps and appeared not to be onerous.
- Date of the lease: The lease is dated 23 June 1987 and therefore is an 'old' lease under the Landlord and Tenant (Covenants) Act 1995. Accordingly, should the tenant assign the lease, they would still be bound by 'privity of contract' to guarantee the assignee's performance of the covenants within the lease.
- Within the Act: The lease has not been contracted out of the Security of Tenure provisions provided under the Landlord and Tenant Act 1954.

Appendix VIII – Valuation

By reference to the hierarchy of rental evidence, a base rental of £4.30 per sq ft was established for the units 10, 12, and 13. Agreed rental adjustments were then applied to reflect:

- The date of the review.
- The prominence and yardage benefits of the subject unit.
- The size of the unit.
- The 10-year assumed lease term.

The rent was agreed as follows:

Base rent £4.40 per sq ft
Add £0.15 to reflect prominence of the unit

	£
Gross Internal Floor Area 4,343 sq ft @ £ 4.55 per sq ft	19,760
Less £0.10 per sq ft to reflect the assumed 10 year term	434
Less £0.20 per sq ft to reflect a discount for quantum	868
	18,458
	say £18,500 pa

Other appendices

The other appendices were:

Appendix I and II	Location plans.
Appendix III	Photograph of property.
Appendix VII	Hierarchy of rental evidence illustration.

Chapter 8

The APC Interview

The APC final assessment interview lasts for an hour and comprises a five-minute introduction, a 10-minute presentation by the candidate on their critical analysis, and around 45 minutes of interview questioning.

Tales of interview difficulties

Although the occasional stories told by some candidates can give the impression that the APC can be a terrifying ordeal and/or the result a bit of a 'lottery', this is not the case. No one doubts that when success or failure is in the hands of three assessors who may make slightly different judgments, that a result can sometimes go either way. Similarly, candidates may not have a good day and be able to perform well, nor receive their favourite questions.

Surveyors who are involved in graduate training, and who are also RICS APC assessors, will often comment on the disproportionate level of unfair circumstances reported by candidates, as against their own experience of participating in many interviews and observing the assessment process as a whole.

When unfair circumstances are claimed, it is almost always by candidates who feel that they have not done well, and are preparing themselves for an adverse result – and also sometimes hoping to convince managers and colleagues that fault lies with the system rather than themselves.

RICS monitoring of assessors

There are still very occasional instances where the APC system does not operate as intended, and a candidate has not received fair treatment (see the section on 'Whether to appeal' in Chapter 12). However, RICS does monitor assessors, and assessors will also highlight to RICS any instances of poor performance among colleagues. RICS investigates any assessors receiving complaints, and where there is sufficient weight of evidence, assessors may be quietly removed (or otherwise re-trained, depending on the circumstances).

Attitudes of assessors

Assessors who sit on the panels do so in support of RICS, and receive only nominal financial remuneration. They look forward to meeting candidates, and hope they do well. From assessors' discussions that can be overheard at lunch time and also at the end of the day at final assessment interviews, assessors are genuinely disappointed when having to refer candidates.

Interview resilience

Candidates' confidence can be fragile on the day of the interview and they will be rightly nervous. However, some candidates too easily allow small difficulties to affect their overall performance. It is important that candidates focus on the questions being asked, and not on the consequence of previous questions on the chance of success – nor indeed the impact of the overall manner being adopted by assessors.

It is indeed impossible to conclude from the manner of the panel as to whether the interview is progressing well. Where a poor candidate provides incorrect answers, the panel may still just nod and smile. A good candidate may become involved in detailed discussions with an assessor, including some points which go beyond APC level, but are nevertheless covered in order to keep an interview running well.

Getting the best out of candidates

Assessors will look to get the best out of candidates, such as through the opening five-minute introduction which includes straightforward general discussions. However, even at this stage, some candidates can fail to deal with simple questions, and create a poor impression to assessors, and greater anxiety within themselves.

An example is, an assessor saying to a candidate, 'Can you just remind us of your optional competencies, and the sort of work you have been undertaking in each?' Most candidates, having recorded experience against their optional competencies throughout their training period can recall the competency headings with ease, and provide articulate responses which help raise their confidence for the more detailed questions that will follow later. Candidates unaware of their competencies have clearly not approached their APC training period properly.

Alertness to case work

Another area where candidates can let themselves down is in respect of the poor responses given to questions which relate directly to the case work in which they have been involved – as reported to assessors in the interim and final summaries of progress.

First, candidates should be able to account for their conduct in case work, such as summarising the client's objectives, or explaining the strategy behind the timing of a landlord's section 25 notice at lease renewal.

Second, given that the APC is competency/experience based, candidates also present themselves poorly when they do not have appeared to have reviewed the work they have submitted to the assessors, and from which interview questions will clearly be asked.

Chapter 6 on the critical analysis includes examples of questions which are fundamental to candidates' work, but can often not be picked up, neither as part of writing the critical analysis, or subsequent interview preparation.

Awareness of RICS and industry issues

Lack of awareness of the name of the RICS monthly journal sent to members, or the name of the RICS president, confirms to assessors that candidates are not giving due attention to issues within the profession. Some candidates state the Royal Institute of Chartered Surveyors rather than Institution within their submissions, and risk assessors drawing adverse conclusions as to a candidate's professionalism.

Assessor frustration

Where there are any reported instances of apparent assessor frustration, candidates need to reflect on any causes. Examples could include candidates not complying with RICS APC requirements, and poor-quality written submissions, both of which can suggest a lack of application.

Candidates may irritate assessors by deliberately providing long-winded answers in order to use up time and hopefully limit the number of questions being asked. Assessors can become frustrated when candidates do not listen to interview questions properly, and have to steer back the candidate from their off-track answers. Some candidates deliberately talk about a side issue because they are unable to answer the precise question being asked.

The risks associated with leaving 'hooks' in written submissions and the interview day presentation were mentioned previously. Another source of frustration can be candidates sitting the interview when they are clearly not ready.

Being ready to sit, and being well-prepared

Good young surveyors do not find the APC an ordeal. They focus on what they need to know, having regard to their work experience, competencies, essential wider requirements, etc., and attend the interview aware of its format, and how they can get the best out of themselves. Part of this mindset develops from having sought appropriate training support for the final assessment. Another key aspect though, is whether candidates are actually ready for final assessment. As mentioned at various points in *How to Pass the APC*, a key aspect is whether candidates are actually ready for final assessment. Too many are not, and require further experience and complementary learning.

Questioning technique, interview theory, etc.

RICS trains assessors, and produces written guidance which illustrates how a good interview technique can be developed.

Questioning for assessors is however a relatively straightforward process. Interview questions tend to be asked with adequate clarity, and when a difficult question may not have been best worded by an assessor, it is acceptable for the candidate to seek clarification. Assessors will sometimes realise that their question has not been presented as fluently as could be the case, and rephrase the question.

Candidates sometimes report that they were often unclear as to what assessors were getting at – almost to suggest that the style of questioning was at fault. It does however tend to be the weaker candidates who make such comments, really because they do not know the subject well enough to pick up on the points sought by assessors.

Interview questions are not necessarily a level 2 or level 3 question. This is because the assessors' judgment of the candidate's level of attainment will often be in relation to the quality of the responses given, including the depth of technical and applied knowledge. Some questions will, however, be level 3 questions, owing to the more detailed issues to which they relate, and which would not be appropriate for a candidate with only level 2.

Interview insights

The remaining parts of this chapter comprise extracts from previous *Estates Gazette* 'Mainly for Students' articles on the APC, and contributions co-ordinated by Midlands Property Training Centre (which have been used by training providers in seminars, or within training material). These are:

* Overview of interview requirements, and illustrations of questioning.
* Sample interview transcripts.
* Sample report outlining reasons for a candidate being referred.
* Typical candidates' questions and answers relating to the interview.

Chapter 9 then examines presentation technique, and Chapter 11 provides further examples of interview questions, together with some illustrative responses.

Overview of interview requirements, and illustrations of questioning

The following information is assembled from previous *Estates Gazette* 'Mainly for Students' features, slightly updated in line with market changes, and variations to the APC.

For most candidates the interview is the most daunting part of the process of becoming MRICS qualified. Candidates will often be far more nervous than for job interviews, for example. Applicants have chosen to be at a job interview, and can walk away. There is no choice but to attend the APC interview.

Nervousness/preparing for the interview

Nervousness can make some candidates feel sick. There can be feelings of not wanting to be there, with some candidates wondering whether they will be able to get through the interview.

Anxiety and apprehension bring all sorts of factors to mind. The feelings mentioned are not uncommon, and reassurance can be taken from the following.

* Assessors accept that you will be nervous, and will make due allowance where it is felt that anxiety has impeded your performance. The chairman may even meet you outside the room and have a few brief words before introducing you to the panel. Be prepared, for example, to give a brief account of the type of work in which you have been involved since graduating, or of the wider

activities of your firm. The purpose of such simple early questions is to help you settle, so do not think during or after the interview 'why did they ask me that'. And if, for example, you are asked what you were doing during a period of unemployment, do not conclude that time unemployed influences the panel's judgment of your abilities.

- Nerves may have eased slightly by the time the presentation commences, but you are still likely to be nervous. It will not be as easy to present the presentation verbatim, or recall all that you were to talk about under bullet point headings, as it was practising. As you run through the presentation, you should be able to settle yourself, but you may still remain anxious as to the questions that you will be subsequently asked. Try not to become anxious if you feel you have provided poor answers. You are not expected to answer all questions, and you will be stronger in some areas than others.
- Effective planning and preparation, and a familiarity with the format of the interview, will enable you to focus on your performance, rather than be preoccupied with travel arrangements, for example, or whether assessors picked up spelling mistakes in hurried report submissions.
- Nervousness can often be a better disposition than overconfidence. Over-confident, even cocky, candidates may not endear themselves to the panel. Some candidates may be very impressed with their achievements, and consider they are involved in more advanced work than their contemporaries. Remember, for example, that you have had two years' experience compared with the 25 years or more gained by an assessor aged 50. The APC is not the place for clever remarks or attempts at being humorous. There may be light-hearted moments during the interview, but this will usually emanate naturally. Overtly confident candidates, free from nerves, tend to be in the minority, but those more exuberant characters should consider the demeanour most appropriate for a professional interview. Colourful shirts and ostentatious ties have their place, but possibly not at the APC interview.

In ensuring that the day will run smoothly, pay attention to preliminaries which include the following:

- Be aware of the precise location. Do not, for example, assume that Coventry means that the interview will take place at RICS headquarters; it may be a hotel venue. Establish the journey time and any possible travel difficulties. Take money for emergency phone calls or take a mobile phone. Is it worth staying overnight in the area? Remember all the items required for the final day, such as additional log book and professional development records, notes/handouts for the presentation, location details and so on.
- Arrive in plenty of time, to relax and prepare mentally. Wait at another venue if need be. Give yourself time to familiarise yourself with the surroundings once at the location. Go to the toilet, smarten up and focus on the task ahead. You should, of course, dress professionally. Confine luggage to the items required for the interview. You do not want to be anxiously scrambling through a briefcase to find the relevant papers. Bags or coats could be kept in the car, or left elsewhere on the premises (although coats can be left within the room if

need be). It may appear unnecessary to state some of the above, but the nervousness generated by last-minute revision can result in mistakes.

* You will ideally have ensured that the report submissions are of the best possible quality. This creates a good initial impression to the panel and can help increase your confidence. Similarly, a good professional development record is highly beneficial.

The interview environment

It is important that you are aware that the interview may take place in a hotel bedroom. Having visualised yourself in a small office or conference room, the realisation that the interview will take place in a bedroom can be destabilising, if only momentarily. A number of panels will sit on the same day and hotel bedrooms tend to be the most feasible means of providing several small interview rooms. The bed and other furniture will be out of sight, and the occasion will still resemble an office-type interview.

Another possible source of surprise on entering the room could be the presence of another RICS representative in addition to the chairman and his two assessors (although, rarely, there may be only two assessors in total). The RICS monitor could be there to assess the performance of the panel. He or she will take no part in the interview or in the decision-making process, and will be out of direct sight. It may be appropriate to shake the monitor's hand if close by or if introduced, but it would not really be necessary to march across a room and meet the monitor if he or she is clearly adopting a lower profile.

The routine adopted on the day by the panel of assessors, and particularly the chairman, will vary. Some interviews will be conducted in a more informal spirit than others. During the time leading to the final assessment, you should not fear that you will be faced with stern-faced assessors looking to give you a hard time. Assessors undertake their work in support of the profession and are on your side. They want you to succeed, but will, of course, not hesitate to fail you if you are unable to demonstrate an adequate level of knowledge, or if you are unaware of professional ethics/codes of conduct.

Assessors may have a different background to you, for example they may be working in private practice or in the public sector. Candidates sometimes feel that the APC is biased in favour of private-sector candidates. This is not the case. Upon qualification, public-sector candidates could move to the private sector, or could even establish their own practice. As well as being familiar with professional ethics, public sector and other candidates working for in-house teams must view the term 'client' as their employer, and not a concept confined to private practice.

Questioning

To be satisfied that candidates meet the standards required to be a chartered surveyor, their performance at interview will be assessed against a number of criteria.

Responses to questions arising from the presentation need to be logical and clear. A good understanding is required of other elements covered in the critical analysis and also in the summary of progress and other submissions, including the professional development record.

Candidates must have the requisite depth of knowledge in their chosen competencies, and be able to put theory into practice. The introduction of the competency basis of assessment helps assessors to focus on areas of experience, but candidates should still be aware of all-round technical matters, and of issues affecting the profession.

Candidates must be aware of the RICS Rules of Conduct, and of their duty of care to the public, clients and employers. In the case of referred candidates, assessors will consider the extent to which previously identified weaknesses have been addressed.

Within each criterion (since amended by RICS, but similar), candidates will be assessed as being poor, weak, satisfactory or outstanding. Poor and weak constitute a referral, and satisfactory and outstanding, a pass. To ensure that an accurate judgment is made, assessors will often tick to the right or left, or split the line between respective boxes on their marking sheet.

The decision to pass or refer a candidate is based on an overall assessment, including consideration of the critical analysis and other submissions. Candidates must, however, satisfy the panel both in respect of their chosen competency areas and professional ethics. If not, the decision will be a referral.

Common questions

There could be common questions, such as 'what are the differences between the roles of an arbitrator and an independent expert?', 'what are the grounds for opposing a tenant's application for a new tenancy?' or 'what are the six valuation rules for compulsory purchase acquisition?'

However, questions can be asked in different ways, and the depth to which a subject is covered will vary between interviews. This will also reflect the experience of candidates, the chosen competency areas and the level of attainment reached.

Finding an edge

An assessor could ask a relatively simple first question in order to begin the examination of the candidate's knowledge on a certain subject. A satisfactory answer could be followed by a more detailed question. The questioning could result in a relatively in-depth debate and could reach a point where the candidate declares the limit of their expertise, and that they unfortunately do not have an answer.

This is a good example of where you may feel disappointed that you have failed to answer the final question, but the assessors are actually impressed at the depth to which they were able to take the questioning.

The poor candidate, on the other hand, may have given an unimpressive answer, with the assessor concluding that knowledge is weak in this area, and that it would be pointless pushing the matter further. A candidate can mistakenly feel that a good answer has been given because the assessor has moved on to a different subject.

How much do I know?

Some candidates can be unaware of how much they do not know, and give what they think has been a correct or complete answer when in fact it was not. It is often difficult for candidates to judge the standards they need to meet, and whether they have done enough at the interview to succeed.

Different assessors may have alternative ways of asking questions. Although they receive training specifically in respect of the APC, questions may not always be asked with perfect clarity. You can ask for further clarification, or ask for the question to be repeated. It will not necessarily imply that you lack understanding of the issues being raised. The assessor may realise that the question has not perhaps been asked as well as it could be, and does not think any less of your abilities for seeking clarification.

There could, of course, be situations where confusion merely reflects a candidate's ignorance, such as in response to, 'what does *caveat emptor* mean', the candidate looks blank, asks for the question to be repeated, and has clearly never heard the term before.

A wide range of questions

Some assessors will ask longer questions than others. The examples of questions below are relatively straightforward. Scenario-type questions could also be asked, describing a property, together with the landlord's and tenant's circumstances, and asking how either might be advised. A certain scenario could be presented once a relatively simple initial question has been dealt with satisfactorily.

Assessors will not always indicate whether your answer is correct. You may hear 'good', 'thank you', 'yes, that's correct' – assessors will always try to be encouraging and supportive. Assessors may move on, without comment, because you have demonstrated a good understanding. In other situations, they could move on because your understanding is clearly poor.

Assessors could help to extract the information from you. Some interviews will be conducted in a relatively informal spirit that provides the opportunity to have more of a guess at an answer once a cautious 'don't know', or 'I would need to take advice' type answer has been given. Further probing, or the rephrasing of a question by an assessor, can sometimes inspire an answer.

Answers need to be concise. Assessors are aware that there are some candidates who attempt to talk tactically for as long as possible (especially on topics on which they feel knowledgeable) on the basis that, while they are talking, the assessors are not asking questions. However, it is not good practice to irritate assessors. Also, in order to be satisfied that candidates are fit to become chartered surveyors, assessors need to ask a range of questions within the allotted time.

Candidates should also avoid repeating elements of the question back to the assessor, such as in response to, 'what are the main factors currently affecting the UK economy?', replying, 'the main factors which I think would influence the UK economy would be ...'. It is preferable to think in silence, perhaps with a thoughtful nod to show that you understand the question. A sip of water could be taken to give thinking time.

Look as if you are thinking. A complete blank could encourage the assessor to move on when, in fact, you may have been eventually able to provide an answer. On other occasions, you may like to jot down a few words on paper before giving an answer.

Avoid repeatedly using words like 'obviously' and 'basically' which also tend to be used to give thinking time, but can suggest that the issues and answers are not, in fact, very obvious to you at all.

Answers should get straight to the point. If asked a specific question, such as about the ability for landlords to recover rent arrears under 'old' leases in accordance with the Landlord and Tenant (Covenants) Act 1995, the actual question should be answered. Some candidates will launch into a pre-prepared commentary about privity of contract, and talk about authorised guarantee agreements.

Some could even say, 'privity of contract' out loud before answering, making it more obvious that there is a lack of genuine understanding, and that their responses comprise rehearsed statements.

What the assessors are looking for

Similarly, if you are asked: 'I appreciate that you did not examine comparable evidence in great depth in respect of the office property covered in your critical analysis as you were only establishing a tone of values for marketing purposes, but, assuming that you were undertaking a rent review, what are the main elements of the comparable evidence that you would be seeking to establish?' – the assessors are, of course, looking for factors including the location and size of the comparable property, whether there were any inducements such as rent-free periods, particular lease terms and so on. Candidates must avoid rushing to explain what comparable evidence is, and why it is used.

It may seem unnecessary to stress that the actual questions should be answered, but under interview conditions, this is not always easy – especially in areas in which candidates are weaker.

Picking up on points

Some questions will expand on points included in the presentation or the written submissions. For example, 'you mentioned in your presentation that you undertook certain works having regard to the relevant health and safety regulations. Can you please expand on what those regulations were?'. Candidates should be aware of the relevant provisions of the health and safety regulations to which they referred, and not be just vaguely aware that they exist.

Similarly, it could sound impressive when you state that measurements were undertaken 'in accordance with the RICS Code of Measuring Practice', but make sure you know what the code says and, most important, how it applies to properties to which you have referred.

Care should be taken when preparing the summary of experience and critical analysis to not, inadvertently, present assessors with the opportunity to ask questions with which you will struggle. The written submissions have to be a truthful reflection of your work and you cannot leave gaps, but in the critical analysis, and more so the presentation, certain areas can be given more emphasis than others. Also, beware of claims of achievement, such as, 'I consider that I have become relatively accomplished in development work'. By whose standards? After two years' experience?

Important areas

You will not be expected to answer all the questions. Your knowledge of primary areas will clearly need to be of a higher standard than areas in which you have had less, if any, experience, but, as a chartered surveyor, you should still have some basic knowledge.

Primary areas encompass the work you have carried out, chosen competencies, the topic of your critical analysis/presentation and the specialist area advised to RICS. The absence of experience is not always justification for unfamiliarity with certain areas of surveying, especially as the professional development programme can be used to fill in gaps in work experience.

Assessors are looking for candidates to demonstrate the requisite level of professional knowledge and also be able to act with professional integrity. Codes of conduct/professional ethics are an essential part of the APC interview. Professional development is also important, and candidates can expect to be asked about the activities detailed in their submissions.

Where reference is made to the term 'client', the question is just as relevant to surveyors working in the public sector, or for in-house property teams, as it is to surveyors in private practice, as they have their own internal client/employer.

There is a vast range of questions that candidates could face at interview – ranging from a relatively detailed examination of work experience gained, to general questions covering areas in which candidates may not have practical experience but about which, as chartered surveyors, they ought to know something. This should be borne in mind when considering how difficult or relevant some of the examples of questions (see Chapter 11) may be.

Difference from university exams

The interview is not about textbook revision. Assessors are looking for you to demonstrate a practical understanding of surveying, having worked in practice for at least two years.

Professional development/private study will help to complement work experience, but your ability to answer questions most effectively is determined primarily by the work experience gained. It is easier to understand the issues, and to answer the questions, if you have been involved in a rent review, for example, than if you merely read books and articles on the subject.

It is also easier to answer questions such as, 'what are the main differences between a rent review and a lease renewal?'. If lease renewal and rent review have been studied separately, without much lateral thought, it may be difficult to provide a good-quality answer at the interview, especially if you have never thought about such a comparison before.

Sample interview transcripts

Set out below is an example of how the initial stages of an interview may run.

Chairman: Richard Lancaster?

(Candidate acknowledges)

Good morning, I am Fiona Davidson, your chairman for the day. Come inside and make yourself feel comfortable.

(Candidate walks into room having been met at the door by the chairman. The assessors will meet you at the door and will not just shout, 'Come in'.)

(Candidate shakes the hand of each of the two assessors, who just announce who they are – 'Pleased to meet you' etc.)

Chairman: Please take your jacket off if you wish.

There is water on the table. Please just top yourself up if you run dry.

(Candidate takes papers out of brief case and places them on the table.)

Settled? OK then.

Have you had far to come today?

Candidate: I travelled up last night actually. It is about a hour and a half's journey from Leeds, and I thought it was best to be here, and settled for a 9.30 start.

Chairman: How did you find the hotel?

Candidate: Fine thanks.

(This largely irrelevant questioning is to allow candidates to settle.)

Other questions could include:

I see you work for Turners – is Dave Unwin still there or has he retired?

I saw from your papers that you took a year out. What did you do?

Is there anything particularly interesting that you have been working on since you submitted your papers?

Again, inconsequential questions allowing the candidate to settle, and not, for example, making a judgment about what the candidate did in their year out.

Chairman: I will ask you to start your presentation in a few minutes time, but just first to introduce your two assessors.

(Names and areas of practice advised for assessors.)

(Note that sometimes there are two assessors in total rather than three.)

(Note that there may be a RICS monitor in the room, taking the total number of RICS representatives to four. The monitor assesses the panel not the candidate, and should not be a distraction.)

Chairman: I am sure that you know what today is all about, but just to outline the process –

following your presentation, which should last 10 minutes, I will have a few initial questions, and will then pass you over to ... (whichever assessor ... – chairman outlines the routine as described earlier in the chapter).

It is not expected that you will answer all the questions put to you. If you do not know an answers, just say so and we will move on. Also, if any points are unclear, please do not hesitate to ask for clarification.

Chairman: Now just before I forget, have you any papers you wish to hand over – log-book, professional development record etc.

(Candidate hands over papers.)

Chairman: Are you fit and well, and ready to proceed?

(This can strike candidates as an odd question, but the chairman is required to give candidates the opportunity to advise if there is anything that needs to be made known. In any event, candidates suffering any adversities or ailments are likely to have notified RICS in advance.)

(Candidate proceeds with presentation.)

The main parts of the interview will be a routine question, answer discussion format.

Sample report outlining reasons for a candidate being referred

The following report is a sample prepared for illustrative training purposes, followed by additional comments. This is similar to RICS format, but is not an actual RICS referral report.

The panel regrets to inform you that you have been referred.

Training and experience

The summary of progress did not demonstrate a sufficient breadth and depth of experience. Your report writing skills also require improvement.

Before seeking to have a particular competency signed off, your supervisor and counsellor should be satisfied that you have attained the appropriate level as per the APC guides.

You are also advised to improve your knowledge and awareness of issues affecting the profession by attendance at relevant training events.

Critical analysis

The standard of report writing skills was poor, with the many errors also giving the appearance that the report was rushed.

Many of the issues covered were not to the required depth and, overall, the report was too descriptive, and not focused on the tasks you undertook personally. It was difficult to distinguish your personal role from that of colleagues also involved in the case.

Interview and presentation

Your presentation was delivered well, but was weak in content, reflecting the points made above.

You did not demonstrate to the panel that you had the required level of understanding of some of the elements to which you referred in the critical analysis.

During the interview the following deficiencies were identified.

- You had not given regard to alternative uses that may be suitable for the property.
- You were unaware of the rateable value of the property.
- You were unaware of the nature of the user provisions in the lease, and of their importance in determining rental value.
- You did not appreciate the distinction between market value and worth.
- You were unaware of the main heads of terms for a compulsory purchase compensation claim, despite commenting in your summary of experience that you had negotiated a number of settlements.
- You were not aware of the constituents of a DRC valuation.
- The panel were not satisfied with your responses to questions on conflicts of interest and the position in respect of fee quotes for instructions.
- Vague responses were given when questioned about professional development initiatives that you had undertaken.
- You did not demonstrate the required level of knowledge in the core competency of valuation.
- Your knowledge and understanding was generally weak. You should look to extend the range of work in which you are involved. The panel suggest that you speak with your Regional Training Adviser.

Reading between the lines of the above report, candidates do not do themselves favours when they:

- Obtain supervisor's and counsellor's certification that they have met the required standard when clearly they have not.
- Choose a project for the critical analysis in which they have not had an active personal involvement, on the naïve assumption that assessors will be impressed with the project instead of focusing on the candidate's experience, involvement and understanding.
- Are unaware of fundamental issues affecting their project.
- Do not do justice to the effort involved in preparing report submissions, and give the impression that they consider attention to detail, presentation, grammar, etc. unnecessary.

- Overstate the extent of the experience gained.
- Cheat professional development responsibilities.
- Assume that assessors are not alert to most of the tricks played by some candidates.

The comments from the panel are inevitably selective, and based on the deficiencies discovered in an interview of only one hour. There may be many aspects that require a better understanding.

Typical questions and answers relating to the interview

Below is a sample of APC candidates' questions and the training provider's responses relating to the APC interview.

Does it make a difference which venue you sit/are some venues easier than others?

If the venue does make a difference, it is suggested, respectfully, that you ought to be more positive, and raise your standard to a level where the venue or any other minor factor do not make a difference. Candidates will normally choose their nearest venue, and except for holidays or other commitments, there is often little reason for adopting a different venue. One situation, however, where it may be necessary to choose a different venue is where candidates are tight on the number of days' experience they require in order to be eligible for final assessment. An extra six weeks or more may be gained by choosing a different venue.

Also, for some routes/specialisms, not all venues are available, and it may indeed be necessary to travel to the Heathrow venue, for example, for your interview. In response to the question, it should be noted also that some assessors travel all round the country to assess, so a London based assessor could, for example, appear on a Harrogate panel.

When do the assessors read the reports?

They are likely to have a quick look on receiving the submissions from RICS (about four weeks before the interview) to check that everything is there. In the week before the interview, they will go through the submissions in detail. The chairman generally has more to do than the assessors. It is worth noting that if you have made mistakes with competencies, there is not a facility within RICS to pick these up and advise you. It is your responsibility to meet the requirements, and if you fail to do so, this may only materialise on the day.

Who are the assessors?

Simply surveyors working in the candidate's route/areas of experience. They have a genuine interest in what they do and are not out to give candidates a hard time. They wish to see you succeed, but cannot be blamed for occasionally appearing impatient with candidates who fall substantially short of the standards required, and, for example, have submitted careless written reports, and just really had a speculative punt at passing the APC.

What should I take into the interview?

It is best just to take the papers you require. Leave coats and bags in the car, or if you have not driven, leave them with the RICS staff, or with someone else at the venue, such as a porter.

Do the assessors indicate if you have got the answers correct?

No. They will generally nod and smile in encouragement in order to help you maintain confidence, but they will rarely tell you that an answer is incorrect. They may deal with incorrect answers more subtly, such as quickly moving to another subject when it is clear that the candidate has little knowledge of the areas being asked. Some candidates consider that they have done well at the interview because all of their answers were met without being declared wrong by assessors. The answers may, in fact, have been poor.

What do you do if the assessor considers that they are right when you as the candidate know that they are wrong?

First, do not argue with assessors. Second, what you may think is right may not, in fact, actually be so. You may also, for example, have misinterpreted the question. Third, there is more to lose by disputing whether an answer is correct, than letting the matter pass – especially as the assessor and, if not, the other two assessors, may have realised their error.

What do I do if I am really struggling during the presentation?

The best thing to do is pause and take a sip of water. The assessors are aware that the presentation can be a particularly difficult challenge for candidates, especially when they are nervous. A small pause may not actually go noticed by assessors, but a longer pause is likely to draw their attention – especially if they were looking down, making notes when you initially faltered. Find your place in your notes/ cards, and even apologise before carrying on.

Do the assessors stop you dead on 10 minutes if your presentation is overrunning?

It would be highly unusual for candidates to be stopped dead on 10 minutes. The chairman would usually warn a candidate at, say 9.5 minutes if it appeared that they were going to exceed the 10 minutes. If a candidate is talking about their reflective analysis/conclusions at around 9.5 minutes, the chairman may just allow them to run out time. If there seem to be several areas yet to be covered, as illustrated, for example, by the headings on a handout, the chairman may give a warning. The candidate would have to rather hurriedly bring the presentation to a close, perhaps by moving straight to the reflective analysis element. If the candidate overruns 10 minutes and fails to complete the presentation within a reasonable additional time, they may be stopped dead by the chairman who then invites questioning. Allowance is, of course, made for nerves, which can cause presentations to last longer than planned. Candidates may, however, not receive the benefit of any doubt if their presentations are set to overrun because of a lack of effort in respect of their preparation. As a general point, assessors' attitudes and approaches vary, and there are no absolutes – unlike the way some candidates as part of their APC preparation search for a black and white image of what will happen, how assessors will make decisions etc.

What do you do if you just do not know an answer in the interview?

The key requirement here is not to guess – thus risking giving incorrect answers and/or appearing reckless. A 'don't know' answer is not ideal either. If possible, respond on the lines of, 'I am sorry, but I have not had any experience in that area. I would however, approach my manager for assistance. The sort of queries I would have would be...'. This turns a don't know response in to something more positive, although the response must still be relevant. Candidates will themselves, have to judge the appropriate response. Not knowing something fundamental in one of the main competency areas, for example, would be more of a downfall than simply not knowing anything about a general matter raised by one of the assessors.

What do you do if you know someone on the panel?

An assessor who knows you should really have declared this to RICS beforehand if it is expected that there may be a conflict, or if it is considered unfair for you to be interviewed by them. Because someone knows you does not bar them from sitting. University lecturers cannot remember all of their previous students, and senior managers cannot recall all the junior staff who have passed under them – and may recognise you only on the day of the interview. This does not have any significance in terms of the questions asked or the result. If an assessor sits on a panel and realises that they should have perhaps declared a conflict, it will be one of the situations where they should afford you the benefit of any doubt.

What do you do if you see your assessor in the toilet after the interview, or in the car park?

Ignore them as best you can, but just politely acknowledge their presence with half a smile if needing to avoid appearing ignorant – but do not open conversation.

Presentation Techniques

This chapter has been prepared in conjunction with Scott Kind, GVA Grimley's training and development manager.

As part of candidates' APC training period, they should begin to gain experience in delivering business presentations. At the APC interview, candidates are required to deliver a 10-minute presentation on their critical analysis. This chapter therefore first provides an extract from a previously published *Estates Gazette* article from Scott Kind on presentation techniques. This is followed by guidance specifically for the APC, including extracts from a previous *Estates Gazette* 'Mainly for Students' article.

Business presentations and general issues

Presentations take different forms, depending on their purpose. A graduate interview might include a five-minute presentation on university work, or a project considered as part of a group team working sessions.

Senior personnel within the large surveying practices may have a short amount of time to make the right impact on a government or major corporate client in order, hopefully, to win new business.

Contributors to success

Key contributors to delivering a presentation successfully include:

- Eye contact with the audience, and avoiding an over reliance on notes.
- Natural rather than over-rehearsed delivery.
- Good structure, comprising an effective introduction, a main section and an ending.
- Appropriate content, having regard to the purpose of the presentation and the nature of the audience.
- The right accompaniments, if any, such as a laptop/projector ('PowerPoint') presentation, handout outlines of the structure of a talk (perhaps including a facility to make notes), a report submitted in advance, or full notes (such as a copy of a speech or separate notes as may be common with CPD events).
- The right tone of voice and speed of speech, and pausing at the right moments in line with the structure of the presentation.
- The demonstration of expertise in the subject, and also enthusiasm and commitment.

Preparation and practise is vital. This will involve setting out the structure of the talk, and being familiar with the detailed content. Practise helps the presentation to be delivered more fluently and articulately on the day, instead of having to formulate too many thoughts as the presentation progresses. It will also help ensure that an allotted time is kept to.

Nervousness

Confidence in the subject area, and in the capability to deliver a good presentation, helps to ease nerves. Experienced speakers will have already broken the nerves barrier, but will still be on edge (and indeed, need to be in order to find the adrenaline to help give their best), albeit with the comfort that they will not falter once on their feet.

For less experienced speakers, including young surveyors, nerves can be intense. Surveys indicate that the aspect of life that most people fear greatest is speaking in front of others. Anxiety can build over the days leading to the presentation, and the impending ordeal can be difficult to shake from the mind. Physical symptoms can include feeling unwell, and also being sick. Here, nerves relate primarily to the fear of the unknown, such as whether nerves can be overcome to be able to actually go through with the task, and what might be the consequences of things going badly wrong.

Young surveyors' presentations will typically be to a small panel of people, such as a job interview or as part of APC final assessment, or to groups of around 10 to 50 people. Nervousness for interview presentations relates more to the importance of the occasion, and the anxiety to get the right result, but confidence in respect of delivering a presentation will still help ease overall nerves.

Group sizes

For small groups, such as 10 to 15 people, the eye contact is personal, and can be distracting. For groups of 20 to 30, it is easier to scan the whole room with eye contact, and when speaking to groups of 50 or more, the intensity of the audience's eyes feels considerably less – and there is also less chance of being disturbed by the body language of someone in the audience who would rather not be there, or disagrees with aspects of the presentation. Larger groups, though, can still unfortunately make the pressure feel greater.

Starting a presentation

At the height of nervousness, which will be as the presentation commences, the audience will usually be the most intently focused on the speaker than they will be throughout the presentation. Once 30 to 45 seconds have passed, some of audience will be making notes, therefore lessening the intensity.

In arriving early, a feel can be gained of the venue, taking the opportunity to stand at the spot from which the presentation will be delivered, thus helping gain confidence. Meeting and talking to some of the audience will also help settle nerves.

The first few comments need to be relatively easy to deliver. Particular familiarity with the opening minute of the presentation helps a presenter quickly break through any nerves, and feel relaxed. Some presenters do this by telling a story about their journey, the last time they visited the venue, or a joke, but this will not often be appropriate for business presentations – especially when the audience is looking to be informed in the limited time available, rather than listen to irrelevant tales. Also, if any such ice-breakers fail to receive the desired response from an audience, it can be difficult for a presenter to recover.

Structuring the presentation

In order to begin a presentation in a way that creates a good initial impression on an audience, and at the same time enable the presenter to ease in, an introduction needs to outline aspects such as the purpose, objectives and content of the presentation.

The main part of the presentation will comprise a number of sections. How many, depends on the length and purpose of the presentation. Too many sections will disrupt the flow of the presentation, with changing PowerPoint slides, other overheads or the audience's notes restricting the impact that a presenter can make by presenting detailed, articulate comment to them.

The end of each section provides an opportunity to take a short pause, check timings, take a sip of water and so on, and also regain the attention of the whole audience. This would work well for a section of crucial importance to the whole presentation; opened in an enhanced way to those already presented. Some presenters prefer notes to be provided to the audience after an event in order to avoid the noise of paper being turned. Others however find that the audience's turning of paper provides greater time to pause than would otherwise be possible; in which case it is important to ensure that page breaks correspond suitably with the presentation.

The availability of a glass of water is an important tool to guard against the delivery of the presentation faltering. If a presenter has stumbled over words, or looks down at their notes and cannot quite recall what to say, a sip of water can provide the vital five seconds or more to regain composure. To the audience, the presenter is having an innocuous sip of water, whereas an undue pause could otherwise indicate a problem, with all eyes suddenly focusing on the presenter, wondering what is wrong and what the next step might be. An experienced presenter knows that even if they have lost their way with the notes, they can pause and remain composed. The alternative is to panic and freeze. Other signals to the audience at a difficult moment, to ensure that all is well, include being seen making a quick note of something, or actually saying, 'that has just reminded me of another point to raise later'.

Sections of a presentation need to correspond effectively with a PowerPoint presentation, or any material handed to the audience. In particular, headings need to be appropriate. Some presenters announce that they are moving on to a particular section, while others have the natural technique to simply state the heading. It can also work well to stress the words of a heading within the first sentence of the section. If anyone in the audience has lost their way in the presentation, this enables them to refocus.

Presenters will need to ensure that the content of the talk is at the right level for the audience; really a balance between not stating the obvious and ensuring that knowledge of the audience is not assumed, and avoiding the talk going over their heads. It may be necessary to liaise with other speakers, if possible, on the content of their presentation, so as not to repeat too many points, or even contradict other speakers. A presenter can come across very well when they are able to bring aspects covered by a previous presenter into their talk by way of brief reference, or an illustrative example. Illustrations always help a presentation.

It also needs to be established how, if at all, an introduction will be made, such as by the chairman for a day's conference, or a colleague handing over to the next speaker. In deciding what will be said, or in providing suitable notes to a host, the start of the presentation can be tailored accordingly on the day.

The presenter's own material

As indicated above, presentations will be approached differently depending on their purpose. While thorough preparation is vital, including the preparation of a full script if necessary, it is important that the delivery of the presentation is natural. An over-reliance on notes will restrict eye contact with the audience, and can also make the presentation sound as if it is being read.

Experienced presenters such as newsreaders and politicians are able to read a full text without it sounding read out – but a full text will not be appropriate for presentations delivered by most surveyors.

Bullet points and additional notes are favoured by most presenters, with regard being given to the information seen by the audience on a PowerPoint presentation or separate notes. A presenter familiar with their subject should be able to look at key words, and deliver the presentation naturally. However, one important aspect of delivering presentations is that the presenter adopts a format with which they feel most comfortable, thus maximising confidence.

Bullet points could appear within a full text, highlighted in bold, colour or larger font size for example. Here, the presenter lifts their eyes off the highlighted area, without being drawn into reading out the text, and can deliver the presentation naturally. The surrounding text is available if need be, but in being familiar with the text through re-reading the text regularly as part of preparation, the presenter tends to be able to easily keep to the planned talk with minimal reference to the text.

Use of PowerPoint

Many presentations are undertaken with the support of PowerPoint. This can be effective if used appropriately, but can sometimes give the impression that presenters are hiding behind the graphics, switching their eyes between screen and audience, and not directly addressing the audience – thus losing the ability to best sell themselves.

Effective inclusions within a PowerPoint presentation include the company logo and attractive backdrop design, title of the talk, name and position of the speaker, and an outline of the areas to be covered. The amount of text accompanying the main sections of the presentation needs to be limited, but content will vary depending on the nature of the presentation, and also whether notes are separately provided, or whether the audience wish to make notes during the presentation (in which case text remaining on screen for some time can be helpful). Photographs and other illustrations are an effective element of a PowerPoint presentation.

At the end of a presentation, the final slide can remain open, including through any question and answer session. The company logo, and presenter's name – and even contact details – could also appear.

It needs to be remembered that an audience's view as to the calibre of a surveyor will derive mainly from what they said, rather than their use of technology (although fluent use of technology is, of course, important). It also needs to be borne in mind that technical aids can malfunction, leaving the presenter feeling stranded, and the audience unimpressed with the speaker, even though the fault may be beyond their control.

RICS APC presentation

At the final assessment APC interview, candidates are required to deliver a 10-minute presentation on their critical analysis, which is the 3,000 word report on a case in which they have been actively involved.

This will be in front of a panel of three assessors, usually in a hotel room across a table in tight surrounds. Most candidates remain seated, and draw on a triangular easel which contains information in a similar way to the content of a PowerPoint presentation, or provide a handout to the panel. (Candidates have the choice to stand up or remain seated.)

Some candidates just talk from notes, without presentation aids, but if struggling under the nervousness of interview conditions, the impression can be given that they have not taken the trouble to prepare property. If the presentation has clearly been structured, as per a handout for example, assessors are unlikely to take such a view.

Some of the points made above are relevant to the APC presentation, but there are also many differences that can candidates need to be aware of. Various courses provide guidance.

APC presentation issues

This extract from a previous *Estates Gazette* 'Mainly for Students' article covers issues specific to the APC.

Content

First establish the content. This will clearly derive from the issues covered in the critical analysis, but you will still need to be selective about the areas covered – the presentation lasts for only 10 minutes.

As with the critical analysis, judgment, analysis, options, conclusions, results and appraisal are key areas that need to be covered. The descriptive elements of the presentation do not provide the same opportunity to demonstrate your ability, and should therefore be kept to a minimum. Assessors will be looking to see how you have performed while working in practice, the extent to which you can evaluate options and recommend appropriate action.

Some candidates feel that they will gain credit in the summary of experience and critical analysis by covering higher profile/higher value work in which they have been involved, but only in the capacity of supporting a senior colleague. However, it is you and your work that assessors want to hear about. If your role has been only one of providing support to a senior colleague, this is likely to be evident to assessors.

In talking about the work in which you have been actively involved, you will have a greater understanding of the issues, and should be able to answer questions more easily. The presentations of candidates who have not had the most active role in their project tend to contain more descriptive detail, and show a lack of understanding.

The presentation tests communication skills, not your use of technology. Particularly for the general practice interviews, it is not appropriate to bring in tripods/flip-boards or use computers.

Handouts

Material is often confined to a handout of up to four sheets for each member of the panel. Assessors will not be able to read text easily while following your presentation, so it needs to be kept to a minimum. A single sheet could outline the main headings.

Plans and photographs can be useful. You may place a plan or photo on the table and point things out to assessors (but remember that you will be on the opposite side of the table and do not want to be twisting your head round to view the material).

While handouts can be a nice distraction from fixed eye contact with the assessors (if they are not taking notes), props can be a hindrance. Some candidates attend with a foldover triangle-type display which has details of their presentation. This is useful if it allows you to demonstrate key information as an alternative to handouts, but you should avoid troubling yourself with the need continually to revolve the notes in order to keep up with the flow of your presentation.

You will have the opportunity either to sit or to stand during the presentation. Although most business presentations will be undertaken standing up, candidates usually sit for the APC presentation. Bear in mind that there may be a short distance between you and the panel, and that standing up could result in you looming over them. The panel are most likely to be making notes and may find it uncomfortable to write and then have to make eye contact with you at a relatively high level.

Structure and delivery

Having determined the content of the presentation, it will need to be organised into a suitable format. Reference should have been made to the requirements of the critical analysis listed in the RICS guides. Consideration has to be given to delivery, and the written information that will be used in support. It is rare that candidates work without any notes at all.

One of the most effective means is to work from key headings and subheadings. A natural delivery will be more attractive to the panel than a standard, well-rehearsed speech, and will enable you to demonstrate a greater understanding of the issues. The main headings could, for example, match the headings on the handout.

Headings are an effective way of breaking up the presentation. At the end of one section, you could, for example, pause, say 'the options available' in order to introduce the next section, pause again, and then cover the issues. If the final section is 'Reflective analysis', make sure that it is reflective and not the provision of further information. Terms such as 'reflective analysis', 'conclusions' or 'lessons learned' suggest that the presentation is nearing its end, and that should be the case.

It would not really be appropriate to incorporate subheadings as opportunities to pause and introduce a new element in the way outlined above. Five to 10 main sections should be sufficient. Any more and the breaks will come too frequently and disrupt the flow of the presentation.

It was mentioned above that it can be harder to perform under interview conditions than when practising the presentation. If the presentation is undertaken from key headings/bullet points, a back-up script could be available, if need be. This will increase your confidence. Overall, you should choose an approach with which you feel most comfortable, not forgetting that the style chosen should not infer a lack of effort.

Make eye contact with each member of the panel throughout the interview and do not be put off by note-taking. Although assessors may appear disinterested in what you have to say, they will be taking their task seriously, but not necessarily nodding in appreciation of your performance.

It is not good practice to overrun. Although the presentation should be a maximum of 10 minutes, it would not generally be regarded as a downfall if it took, say, eight or nine minutes. Under interview conditions, the presentation may be given at a different speed than when practising. If headings or bullet points are used as triggers for ideas then time will be taken for the deliberation prior to delivery. If you have to resort to reading a prepared script, the presentation is likely to be delivered at a faster pace.

The chairman may stop you if you exceed 10 minutes, especially if the end of the presentation does not appear imminent. You could also be warned that you have one minute left. Proper planning should ensure that there are no such difficulties. Do not become disheartened if the presentation goes poorly.

Chapter 10

What Candidates Need to Know in Their Subject Areas

This chapter provides illustrations of what candidates need to know in their subject areas. It is important, however, to stress that the APC does not work on the basis of RICS providing a prescriptive list in each competency area of what candidates ought to study. This would be impossible anyway, owing to the many variations between candidates, depending on their individual areas of work experience, optional competencies, choice of critical analysis, specialist area selected for final assessment, and other factors.

Although the RICS guides provide relatively little information on the aspects covered under each competency, and what exactly candidates must understand to reach level 1, 2 or 3, in practice, candidates initially gain the necessary understanding through discussions with colleagues. Also, as candidates develop their experience, the key aspects within subjects in practice, and corresponding APC competencies, become clearer.

The sections within this chapter are as follows:

- Valuation.
- Landlord and tenant.
- Real estate management.
- Purchase, disposal and leasing (property marketing).
- Planning, development appraisal, economic development/regeneration and CPO.
- Local assessment/taxation (rating).
- Other competencies.
- Telecoms.
- Other issues.
- Case law.
- Professional ethics.
- Current issues.

A final section covers particular issues for public sector candidates, although which is suitable for all candidates working in an in-house role. Where input has been provided by contributors, details are given in the appropriate section.

Valuation

The core competency of valuation incorporates the following, although as mentioned above, the emphasis will be different for each candidate, depending on the nature of their experience, other competencies, etc. There are still, however, certain fundamental aspects of valuation which candidates should understand, irrespective of the experience they have gained.

- Rental valuation.
- Investment valuation.
- Development valuation.
- Other valuations.
- Red Book (RICS Appraisal and Valuation Standards).

Rental valuation

All candidates should understand the principal factors affecting rental value. Candidates involved in rent reviews, lease renewals and rating especially, must understand the detailed factors affecting value, as relevant to negotiations in which they have been involved. Pricing for lettings does not tend to be as precise, but an understanding is still required of aspects such as the effect of a break clause or longer or shorter lease on rental value.

Investment valuation

The understanding of investment valuation is a fundamental area of knowledge and understanding required by chartered surveyors. It is, however, an area with which APC candidates can sometimes struggle. As is evident at APC interviews, investment valuation exams undertaken at university can be passed without having a genuine understanding of the subject – rather a proficiency in remembering where things go. In contrast, a lack of genuine understanding will be evident to a panel of three assessors. Candidates need to be able to value under-rented, market rented and over-rented investment profiles, understand the merits of 'term and reversion' and 'hardcore/topslice' (or 'layer'), and be familiar with terms such as all risks yield, initial yield, net initial yield, equivalent yield, and purchasers' costs.

Development valuation

Development valuation and development appraisal are very similar in terms of inputs to calculations, but have fundamental differences despite the inter-changeable terminology often used in practice. A development valuation is a market-based estimate of the price that could be achieved for a property. In contrast, a development appraisal is often undertaken on behalf of a specific developer to consider whether they should seek to acquire a site, and if so, how much they should pay. While regard is still likely to be given to market values, the surveyor has more of an advisory role, and within the calculations, will be including inputs which are specific to the developer – including, for example, cost figures provided by the developer himself.

Development appraisals may be undertaken for other purposes, such as on behalf of a lender wishing to consider a particular developer's plans for a site – but the overall point here is that candidates need to take a suitable approach to development appraisal as a competency. Level 2 may be achieved by undertaking only development valuations (which could be in an agency as well as purely valuation capacity), complemented by suitable professional development, whereas level 3 will need to involve acting on behalf of developers/purchasers and undertaking genuine development appraisals.

Other valuations

Even if not gaining experience in practice, candidates should be familiar, for example, with the purpose and basic headings within a development valuation/ development appraisal, a DRC valuation and an insurance valuation/reinstatement cost assessment.

Any valuation experience gained by candidates, such as in respect of specialist property types, or valuations for specific purposes such as capital gains tax, means that detailed knowledge is required of valuation methodology, market issues, any supporting legislation, etc.

Red Book

All candidates should understand the purpose of the Red Book, and be familiar with its principal content. They should also be familiar with any more detailed aspects relating to work in which they have been involved. This could be extensive for candidates working, for example, in the valuation department of a large practice, undertaking valuations for a range of purposes, such as investment and development valuations for loan security, and DRC valuations for company accounts.

Illustration – investment valuation

This illustration of what candidates need to be aware of in respect of in investment valuation is based on a large practice's internal guidance implemented for supervisors and counsellors in their London valuation department, and on an extract from their nationwide in-house training programme. This section provides good examples of how candidates and employers can develop systems which reflect the nature of the experience being gained, having regard also to APC requirements.

Extract from guidance to valuation department

Graduates will join the practice's valuation department either direct from university, or after having gained experience elsewhere (within the practice or externally).

Below is an indication of the competencies, and levels, that graduates should be able to achieve in the valuation department. Some depend on the nature of the work involved, and graduates may have achieved certain levels prior to joining the department.

Valuation (level 3)

A graduate should achieve level 3 if working in the department for six months or more. Some graduates, however, will be at level 3 within three months, such as through a concerted approach to complementary learning/professional development, or because of previous experience.

Inspection (level 2)

Achieved through the usual inspection work.

Measurement (level 1/2)

Level depends on the type of work being undertaken. Experience can be developed in other departments.

Landlord and tenant (level 1/2)

Achieved through familiarity with lease terms, and their effect on rental and capital value, occupation and management. Experience can be developed in other departments.

Purchase, disposal and leasing (level 1)

Achieved through general market awareness, and consideration of market-ability as part of valuation work. Experience can be developed in other departments.

Insurance and risk management/insurance valuation (level 3)

Not a commonly taken competency, and needs to involve detailed valuations rather than straightforward rate per m^2 application. Basic experience is incorporated within the main valuation competency.

Development appraisals (level 2)

Achievable due to the development valuation/development appraisal overlap. Experience can be developed in other departments.

Planning (level 1)

Achieved mainly through planning enquiries.

Investment valuation

The following examples are designed to help supervisors and counsellors determine the level that an individual graduate has reached in investment valuation. The knowledge required in development valuation, and any other types of valuation, depends largely on the type of experience gained, and comment is made later.

Level 1

Five methods of valuation – understanding of the different types of property requiring valuation, and the basic valuation approach adopted.

Factors affecting rental value, investment value, development value – basic understanding.

Red Book – awareness of the purpose of the Red Book.

Level 2

Aware of the processes undertaken at the commencement of a valuation (conflict of interest, and terms of instruction in particular).

Red Book – awareness of the main Practice Statements, and Guidance Notes relating to the work experience gained. Knowledge of the types of valuation that are covered by the Red Book.

Valuation/appraisal – understanding of the distinction, and ability to provide examples of why this is important in practice. Similarly, understanding of market value, worth, price.

Valuation methodology – rack rented and term and reversion valuation – basic knowledge.

Yields – knowledge of the use of yield/multiplier within valuations.

Level 3

Types of investors – understanding of the types of investment property, the types of investors they would be suited to, and how investment appraisal (as well as valuation) would be undertaken.

Red Book – awareness of the practical application of valuation bases (OMV/ERP/ERRP if within the graduate's training period, and MV/special assumptions).

Valuation report – awareness of typical headings, and knowledge of caveats/disclaimers included, and why.

Valuation methodology – working knowledge of rack rented, term and reversion and hardcore/topslice application. Understanding of risks to the investor with over-rented properties, and how to value over-rented properties (including differences with rent review and lease renewal reversion).

Yields – knowledge of yields/terminology – all risks, equivalent, initial/net initial, reversionary, equated. (Interchangeable terminology, and some text book inconsistency can cause difficulties for graduates.)

Purchasers' costs – knowledge of why they may be included or excluded in a valuation.

Vacant properties – knowledge of void costs and how an investment property would be valued if vacant.

Factors affecting investment value – more detailed awareness of covenant strength (including how it may be measured) and lease terms on investment value, among other factors including location, specification, etc.

Quarterly in advance – awareness of its use.

Leasehold valuations – basic awareness of how value derives from a leasehold interest, and how valuations would be undertaken.

Hope value/marriage value/ransom value – able to provide examples, and comment on how a Red Book valuation would, or would not, reflect such value.

Specialising in valuation

Some candidates select Valuation as their 'specialist area' for final assessment. If doing so, graduates' knowledge is expected to be higher than if specialising in another area, and includes:

More in-depth awareness of the investment market.

Quarterly in advance – understanding of its application.

Growth explicit valuation techniques – awareness of the application (although actual experience is not essential).

IRR, running yield, exit yield, etc. – awareness of application.

More detailed knowledge of the Red Book and general valuation principles, including DRC, than would be the case for most graduates if those areas had not been selected.

Working knowledge of leasehold valuations.

An awareness of the purpose and format of CGT, CPO valuations.

Rental valuation

Detailed experience in rental valuation will derive mainly from rent review, lease renewal rating and lettings work.

Development valuation/development appraisal

As mentioned above, the extent of knowledge required in development valuation/appraisal and other valuations, including specialist property types, depends the level of experience that graduates have gained.

As a broad indication for development valuation (which would be the competency of 'Development Appraisals'):

Level 1

Factors affecting development value – basic awareness.

Planning – basic awareness of the planning system, the range of alternative uses that may be feasible for a site, and the planning issues that would be relevant. (This is similar to Planning at level 1, but graduates may not have taken Planning as a competency.)

Residual valuation/appraisal – awareness of the straightforward format, and inputs of GDV, Build costs, Developer's profit, Residual value.

Level 2

Familiarity with the inputs of development valuation/appraisal, for both commercial and residential – building on the main headings above. Knowledge of the application of DCF.

Level 3

Level 3 (as mentioned above, not achieved only by undertaking valuations, but rather achieved through acquisition/development work also).

Understanding of the sensitivities of development valuation/appraisal including the variables that can be particularly sensitive.

Understanding of factors such as the effects of a pre-let on an appraisal, how to calculate finance costs, the profit percentage developers require, purpose and level of contingency for particular types of development, how remediation/ abnormals/planning gain/social housing, etc., might be accounted for in a valuation/appraisal.

Understanding the effects of changes in market value over the build period on the developer's profit or loss.

Understanding of DCF approaches.

Familiarity with software used for development valuation/appraisal.

Ability to relate physical on-site factors and planning issues to development valuations/appraisals undertaken.

Other valuation – meeting competency levels

Graduates involved in areas other than investment and development valuation need to understand the valuation methodology involved, and the market characteristics of the particular property.

In their written submissions, graduates should advise whether they have had only an insight, or detailed experience in a particular area (as this will determine the level of interview questioning).

Supervisors and counsellors will be signing graduates off in relation to their proficiency in investment and/or development valuation, and a judgment in respect of levels attained in a specialist valuation area will relate more to interview preparation, than to the signing off of competency levels.

Extract from internal graduate training programme

The main valuation task concerns the valuation for loan security purposes of the headquarters office in London (although areas, tenants, etc. are adjusted). For the practice's other offices, different figures will be provided within supplementary notes, but the requirements will be broadly the same. The tenants and rents are stated for the training exercise, and are not actual tenants or rents.

Illustrative areas and figures are provided so as not to mislead or create confusion with the figures for the building itself, and also to enable valuations to be undertaken which raise a number of issues.

The exercise involves working alone for some of the tasks and in groups for others. Tasks require allocating between candidates, depending on how many are in the group.

Information is as follows:

Area:	100,000 sq ft
Basis:	Multi-let
Management:	Efficient service charge arrangements are in place
Tenants:	20,000 sq ft (The practice)
	30,000 sq ft HM Government (on 2 leases)
	12,500 sq ft (International accountancy practice)
	5,000 sq ft Amco – Amercian internet start up company
	12,500 sq ft (Serviced office provider)
	20,000 sq ft Void

Rental profile/ lease terms:	(The practice)	20-year lease, 17 unexpired, no breaks rent review 2 years' time. £1m pa
	HM Government	15,000 sq ft – 25 year lease, 12 unexpired, rent review 3 years' time, break at five-yearly intervals. £925,000pa
	15,000 sq ft	20 year lease, 12 unexpired, rent review 3 years' time, no break. £900,000pa
	(Accountants)	10 year lease, 9 years remaining. No breaks. 1 year rent-free period. £1.125m pa
	Amco	10 year lease, just granted, £400,000 pa
	(Serviced offices)	10 year lease, 8 remaining, break at 6 months' notice at any time. £800,000pa
	Void	(Available for immediate occupation, good specification and condition)
Market rent/ ERV:		(One of the group is to establish this by speaking to the practice's Agency department, making contact with a fellow APC candidate first. Obtain focus/feature coverage from *Estates Gazette* and *Property Week*, and consider any other sources of comparable evidence, including walking the area and identifying boards. It is not necessary to speak to all other agents, but it would be helpful to obtain one set of another agent's marketing particulars).
Market yields:		Obtained similarly to above.
Measurement:		One of the group needs to circulate the *RICS Code of Measuring Practice* (5th ed) to the others. A measurement needs to be undertaken of the practice's offices occupied by the L&T department only, together with sketch plan.

The valuation is to take place in respect of the individual areas/tenants (effectively assuming they are a single investment), but also of the whole. This allows the individual covenant strength, circumstances, valuation, etc., issues to be examined. Only a final total figure need appear in the valuation report. Calculations are to be provided separately in an appendix entitled 'Calculations/training notes'.

A copy is attached of a valuation report prepared by the practice for another property. This provides a format for the valuation to follow.

Candidates are each required to report to London City bank on the open market value of the property, and comment on its suitability for lending purposes. (Only Market Value is required, but see other reports for detail on valuations for vacant possession, exit value, etc.)

Each candidate is to prepare their own valuation report, drawing on information collectively obtained, but undertaking the valuation and full report themselves. Standard caveats need to be included (as contained in the sample report shown).

The reported information will obviously not be as comprehensive as that contained in the sample report, but headings can be inserted, followed by suitable comment (e.g. 'lease terms – all leases not examined'). Rateable value information will not be available, but an indication could still be sought (see *Estates Gazette* 'Mainly for Students' 15 April 2000 for a general overview – this is also another example of an area which will be picked up in training).

Additional tasks

What marketing strategy would you recommend for the void area – type of tenant, advertising, particulars, lease terms, incentives etc.?

What are the disadvantages between single let and multi-let?

What might rent review strategies be (noting the void, and the other profile/ periods to rent review)?

Examine a copy of the actual lease held by the practice, and report to (trainer), this time as prospective tenant, on its main terms, including assignment and subletting, break, user, rent review basis, repair liability, and service charge arrangements (and anything else of note highlighted in the lease).

The valuation above concerned an office investment in the City of London. What other factors might have been relevant in the case of:

- Oxford Street retail investment.
- Edge of London industrial estate, built 1970s.
- Secondary offices, only half-let.

Examine the *RICS Code of Measuring Practice* in respect of retail and industrial premises, in addition to offices as part of the case above.

Landlord and tenant

Landlord and tenant comprises rent review and lease renewal work primarily, but also a wider understanding of aspects relating to the landlord and tenant relationship.

This includes the types of agreement which may be granted: leases, licences, tenancies, tenancy at will, etc., what is meant by contracting out, and which are the key lease terms considered by landlords and tenants when property is initially let.

Candidates must also understand how lease terms effect the ongoing estate management of a property, incorporating rent collection, tenants' compliance with covenants, service charge administration, dilapidations/disrepair, sub-lettings and assignments, insurance, buildings/facilities management, and health and safety issues. All these areas fall under the competency of Real Estate Management in terms of their implementation in practice, but the candidate with the landlord and tenant competency must still be aware of the fundamental issues. An example would be knowing the basis upon which damages for disrepair would be calculated at lease expiry (cost or diminution in value, which ever the lower – section 18(1) Landlord and Tenant Act 1927).

Candidates should also understand how lease terms affect rental value and investment value. It is particularly important that candidates with the landlord and tenant competency understand the more detailed factors affecting value, such as the effect of a particular unexpired term, the allowance made for a restrictive user clause, how rent-free period incentives can be devalued, and the adjustments made for differing rent review frequency.

As an example of how the requirements of candidates depend on their areas of experience, a candidate may not have the competency of Landlord and Tenant, but must still understand the lease terms which have an affect on rental and capital value if undertaking marketing, investment, or valuation work, and also if involved in development work, and having to secure vacant possession from tenants.

For candidates involved in arbitration and independent expert work at rent review, and/or court processes and PACT at lease renewal, the competency of Conflict Avoidance, Management and Dispute Resolution Procedures may be appropriate. This is also a mandatory competency at level 1, but can be brought in to level 2 or 3. Although APC candidates will rarely be named on submissions to third parties, they may have prepared the majority of the submissions, and generally gained a good insight into the process.

Set out below, is a list of key points associated with lease renewal. This is provided so that candidates are able to develop similar lists for themselves in their competency/subject areas, having regard to their circumstances (competencies, subject of critical analysis, specialist area etc.). The list is only illustrative, noting the many variations between candidates as to what they need to know. Candidates should be able to add further points, and may also establish a further list of 'rental valuation/analysing rental evidence' which covers a range of factors affecting value; some general and others specific to individual property types. A further list could also be created under the heading of 'Dispute resolution for L&T', and cover more detailed issues relating to arbitration and independent expert.

- Opportunities/options for landlords and tenants at lease renewal.
- Bringing a lease to an end – s25, s26, s27.
- Validity of notices.
- Statutory continuation (s24) and interim continuation (s64).
- Business occupation (s23).
- Competent landlord.
- Section 40 notices.
- Valuation date for rent/date of commencement of new tenancy.

- Interim rent.
- Tactics at lease renewal/regarding notices.
- Terms of the new lease: holding (s32), duration (s33), rent (s34), other terms (s35) and how these are established.
- Court processes, and PACT.
- Grounds for possession, and requirements under each.
- Compensation for disturbance and compensation for improvements.

Real estate management

As indicated above, the real estate management competency incorporates areas such as rent collection, tenants' compliance with covenants, service charge administration, dilapidations/disrepair, sub-lettings and assignments, insurance, buildings/facilities management, and health and safety issues.

It is very unusual that candidates would gain experience throughout all these areas, in contrast with landlord and tenant, for example, where candidates tend to gain experience in both rent review and lease renewal relatively easily.

Purchase, disposal and leasing (property marketing)

The purchase, disposal and leasing competency incorporates lettings on behalf of a landlord/owner, leasehold acquisitions on behalf of a tenant/occupier, sales and purchases.

Not all such areas of marketing work may be undertaken in practice, such as where a candidate works with an in-house property team who have a role of either landlord or tenant, but not both. Assessors will reflect the nature of candidates' work in their interview questioning, but still expect candidates to comment on the objectives, aspirations etc. of the other side, and be able to explain how they would approach a particular scenario. Less detailed knowledge and practical understanding would be expected, compared with questioning in relation to specific case work referred to by candidates in their written submissions.

Candidates with the purchase, disposal and leasing competency must show to assessors that they are more than deal-makers, and have the requisite knowledge on matters such as market conditions/market overview, the impact of lease terms on lettings, how lease terms and tenant selection influences investment/capital value and how a landlord may be able to protect future aspirations to redevelop or refurbish.

It is also important that candidates understand agency law. They should be familiar with the Property Misdescriptions Act 1991, Misrepresentation Act 1967 and Estate Agents Act 1979, and also any other aspects – noting for example the impact of data protection legislation, the Disability Discrimination Act and money laundering regulations on an agent's work. Candidates should also be familiar with the RICS Estate Agency Manual. Agency law does, of course, link with professional ethics (see below).

Planning and development

How to Pass the APC is for general practice surveyors, although there are a small number of candidates who may take the Planning and Development route, hence the comments in particular in Chapter 2, regarding this and the Commercial Property route. Candidates in the Commercial Property route may take optional competencies of Planning, Development Appraisals, Economic Development (Regeneration) and Compulsory Acquisition and Compensation (CPO).

To further illustrate interview issues, the comments below are extracted from a publication 'Planning, development appraisal, economic development and CPO' by GP Property Education as part of their *APC Questions and Guidance* series.

Candidates in the APC route of Planning and Development will see their interview orientated to planning and development issues. Here, for example, if a candidate has Valuation as core competency to level 2 (which is possible in the P&D route), only a basic knowledge of investment valuation would be needed. It might not be appropriate for a candidate with relatively little pure planning experience to take the planning and development route – rather the Commercial Property route. Examples sometimes include candidates in the regeneration sector undertaking development appraisals, land sales and acquisitions, and determining regeneration strategy – but not being involved in the securing of planning consent, including through appeals.

Candidates in the Commercial Property route who have undertaken a large amount of planning and development related work may opt for the 'specialist area' of Development and Planning – as stated on final assessment application forms. Some candidates who are involved in planning and development work may actually select the specialist area of Property Marketing – such as because their prime role is the disposal of development land.

The questioning faced by candidates at the APC interview reflects factors such as their chosen competencies, areas of experience, choice of critical analysis, any case examples mentioned in summary of progress/experience etc. As well as preparing for their interview in their competency areas, it is important that candidates are aware of related elements to their work which may not be competencies, but still need to be understood. Examples include:

- Candidates with Planning as a competency still need to know the basics of development appraisal/residual valuation – not least with Commercial Property candidates who will have the core competency of Valuation (which includes development valuation and investment valuation). Consideration of suitable uses for sites will, of course, have to give regard to viability – including occupational and investment demand for end uses, affect of any abnormal costs etc. Although candidates would need to know the basics of residual valuation/development appraisal, assessors would accept that little experience may have been gained in practice, and not, for example, expect candidates to be familiar with matters such as typical developers' profit figures in a certain market, or how a computer package calculates the finance cost.
- Candidates with Development Appraisals as a competency still need to

know the basics of the planning system, despite not having Planning as a competency. Development appraisal is dependent, of course, on factors such as the uses suitable for a site, design/density issues, planning conditions/obligations/section 106 agreements/planning gain, time taken to secure planning etc.

- Planning and development candidates not having the Landlord and Tenant competency (which involves rent reviews and lease renewals as well as a good overall understanding of L&T issues), still need to know the basics of: what rights occupational tenants may have, whether possession can be secured and what compensation might be payable.
- Candidates in the Planning and Development route or in the Commercial Property route specialising in Development and Planning, even if they have not selected the Economic Development competency (which is really regeneration), still need to have a basic knowledge of the UK regeneration sector.
- Similarly, candidates in the Planning and Development route, or in the Commercial Property route specialising in Development and Planning, even if they have not selected the Compulsory Acquisition and Compensation competency, still need a basic awareness of how compulsory purchase powers can be drawn on to deliver planning policy, for example. What candidates need to know about compulsory purchase depends on their circumstances and whether the optional competency of Compulsory Acquisition and Compensation has been taken. Candidates would typically take Compulsory Acquisition and Compensation to level 2 or 3 because they are involved in the negotiation/valuation of claims. Some candidates may, however, have taken Compulsory Acquisition and Compensation as an optional competency because it is a key part of their delivery of regeneration. Here, experience may be gained in the strategic issues relating to land assembly, including whether a CPO can be secured, rather than the detail of valuation.

For the individual competencies, it is important that candidates outline to assessors exactly what the nature of their work is. This will be achieved initially in the written submissions (the summary of progress). At the interview, opportunities to explain to assessors the precise nature of the work being undertaken may arise during the opening five minutes of general discussions/introduction, and also as assessors conduct questioning. It may also be possible to convey to assessors the nature of some experience being gained by suitable comment in the presentation. The presentation is on candidates' critical analysis, and it will not be appropriate for candidates to provide an overview of their background and summary of case work, but it may be possible to quickly convey key points to assessors, more subtly, as part of accounting for why a particular case was chosen for the critical analysis. Assessors will often, of course, be aware of the type of work in which candidates are involved because of the nature of their employer.

Interview questioning will, of course, depend on whether candidates have a particular competency to level 2 or 3. Although the front page (not included in

How to Pass the APC) comments on levels 1, 2, and 3, and that indications will be given in the guides, it is important to bear the above points in mind regarding aspects candidates need to know beyond their competencies. Where levels 1, 2, 3 are provided in the text within this document, this relates to the competency being considered – e.g. Planning, or Development Appraisals in its own right.

Where candidates have selected a competency, but have concentrated on a particular facet of work within that competency, it is, of course, important that, through surrounding CPD/professional development study, that they gain sufficient knowledge and understanding of wider issues relating to their subject. For instance, in the above example regarding compulsory purchase, the candidate involved in the strategic/regeneration aspects would still be required to be familiar with the six valuation rules, heads of claim etc. – not least because such factors determine the overall cost of land assembly – a key point in strategic acquisition/route planning aspects of a scheme. Disturbance compensation, in addition to land values, could be a considerable expense for some businesses/land uses.

Local taxation/assessment (rating)

The competency is termed Local Taxation/Assessment, but is really rating. This involves a general understanding of the rating system, and an emphasis on minimising business rates liability on behalf of occupiers. However, where APC candidates work for the Valuation Office Agency (and equivalents elsewhere), assessors will reflect the nature of their work in interview questioning.

As an indication of the areas to which candidates need to give attention, the following is extracted from GVA Grimley's Annual APC Conference of 2003, and is prepared in conjunction with Claire Paraskeva, an associate in their Leeds office. This is only a basic overview, but illustrates how candidates can develop their own lists of key points relating to their areas of work/competencies.

- Background – basic awareness: General Rate Act 1967. 1973 – the last Revaluation prior to 1990. Local rate poundages. 1990 changes to the rating system. Local Government Finance Act 1988. Revised basis of assessment (FRI). Antecedent Valuation Date. Quinquennial Revaluations. Uniform Business Rate. Transitional Relief.
- Changes for the 2000 Revaluation: Limits on effective dates. Certification. Programming. Target dates.
- Who's Who in rating: Office of the Deputy Prime Minister. Billing Authority. Valuation Office. Valuation Tribunal. Lands Tribunal, Court of Appeal. House of Lords.
- Appeals issues: note expert witness role and basis of fees.
- Definition of Rateable Value: LGFA 1988 – Schedule 6. Rating (Valuation) Act 1999. Economic repair.
- What is a Hereditament?
- Who is the Rateable Occupier?
- *Valuation Date v Material Date.*

- Setting the 'tone'.
- Forms requesting supply of information.
- Hierarchy of evidence. Analysis and valuation of evidence.
- Methods of valuation: Rentals. Profits. Contractors.
- Grounds for appeal: Appeal against the RV (examination of the evidence, floor areas, construction type, inherent defects, state of repair, building date, major disabilities). Material change in circumstances (MCC). Valuation Office Notice. Lands Tribunal or Valuation Tribunal Decision. Mergers/splits. Deletion. Requesting new entry in the List. Change of description/address. Other issues.
- Empty rates: Use of property. Change of Description. Listed buildings. Splits. Other issues.
- Transitional Relief.
- Case law.
- Current issues.

Other competencies

Comment is provided briefly below on other competencies that may occasionally be selected by candidates, but which need due consideration.

Corporate real estate management

Corporate estate management involves working closely with clients to ensure that their property interests are handled effectively, having regard to their operational business interests. Some candidates actually attend the APC interview unable to answer questions such as, 'What is the difference is the difference between the competencies of Real Estate Management and Corporate Real Estate Management?', or 'Which RICS faculty covers this type of work?'. As with all competencies, it is important that candidates understand the components within their competencies/areas of practice.

Corporate recovery and insolvency

If candidates undertake this competency, they must have a detailed understanding of the general roles of receivers, administrators etc., and of the specific roles that surveyors may have, together with knowledge of the relevant legislation, and how it impacts on sales, lettings and management etc. work. Some candidates take this competency because they work in an insolvency department, but if undertaking sales and valuations on the instruction of a colleague, will learn little about the nature of insolvency in a property context.

Insurance and risk management

Candidates undertaking detailed insurance valuations may find this competency to be suitable, but if experience is limited to the basic application of an area figure to a building cost rate per square metre, this will be insufficient.

Real estate finance and funding

This is suitable mainly for general practice candidates involved in investment agency, and relates to factors affecting the ability to raise finance, gearing and portfolio issues etc., and not just an understanding of types of finance and how finance is raised (which will rarely be something in which a graduate/APC candidate is involved).

Telecoms

Telecoms is not a competency, but is available to candidates to select as a specialist area as part of the Commercial Property route.

It is important that telecoms candidates demonstrate sufficient familiarity with non-telecoms property types, even if relatively little experience has been gained in practice. As an example of how this can be achieved through appropriate surrounding training, Crown Castle's four candidates based in Warwick established a programme which resulted in first time passes for all at the Spring 2003 final assessments.

If telecoms candidates are to benefit from the representation of specialist telecoms work as part of the APC, they must possess a strong knowledge of telecoms issues – covering planning issues, technical/technology issues, market developments, code powers, the rating system, key lease terms etc.

Other issues

The purpose of this separate section of 'Other issues' is to stress the importance of candidates understanding areas of work that might not fall under competencies, but which are relevant to their day-to-day work.

For example, a candidate with optional competencies of Landlord and Tenant, Real Estate Management, and Purchase, Disposal and Leasing should still have a basic knowledge of the rating system, the planning system and VAT on rentals and sale prices.

A planning and development surveyor should be familiar with tenants' rights, whether and how possession could be secured, and the extent of compensation payable.

Professional ethics

As stressed by RICS in the APC guides, and as represented by the level 3 mandatory competency in Code of Conduct, Professional Practice and Bye-Laws, professional ethics is an important part of the APC.

Candidates should refer to the RICS Rules of Conduct and accompanying guidance notes. Not all the issues arising at graduate/APC level will be covered, and candidates will need to discuss some areas through with colleagues. An indication of some of the aspects with which candidates need to be familiar, is set out below, again for candidates to add to the list as appropriate. Some aspects will overlap with other areas, such as property marketing and the Purchase, Disposal and Leasing competency, where an understanding of agency law will be important.

- The general content and purpose of the RICS Rules of Conduct and accompanying guidance notes.
- Nine core values.
- Conduct unbefitting a chartered surveyor.
- Conflict of interest – able to provide examples, and consider scenarios.
- Connected persons – able to provide examples, and consider scenarios.
- Terms of business/letter of engagement – requirements, and typical contents for areas of practice.
- PI cover – purpose and principal issues.
- Clients' accounts – purpose and principal issues.
- Complaints procedure – purpose and principal issues.
- Gifts, hospitality – what may and may not be acceptable.
- Not working outside areas of expertise – and awareness of options of sub-instructing others and introducing business to others, and relevant issues/requirements.
- What is meant by secret profits/hidden profits.
- Quoting fees – when fees may reasonably be negotiable, and when fees cannot be cut.
- Rules regarding business names, advertising, logos, use of term 'chartered surveyors' etc.
- Client confidentiality – able to provide examples, and demonstrate importance.
- CPD/lifelong learning requirements.
- Acting outside principal employment.
- Other legislation – awareness of purpose and principal content, with detailed knowledge depending on candidates' other areas of work.

Current issues

Candidates need to be able to demonstrate to assessors that they understand the importance of keeping up to date with developments in legislation, market practice, etc., and are familiar with current issues affecting their work, and the property market in general. It is also important to be familiar with the work of RICS, and any new initiatives which RICS may be pursuing.

The majority of information for graduate/APC level requirements will be reported in the professional journals, which candidates should read on a weekly basis throughout their training period. If not regularly reading the journals, and generally leaving interview preparation until the final few weeks, it can be difficult to be fully appraised of current issues.

Candidates need not learn the precise detail of proposed changes to legislation which may or not take place, and is still speculative to some extent. An awareness of the key features, and how they may relate to the candidate's work would usually suffice. If, however, there is a new version of the Red Book, or a new piece of legislation has been implemented, this will form the basis on which candidates are working in practice, and a greater level of knowledge is therefore required.

With legislation such as Part III of the Disability Discrimination Act 1995, effective from October 2004, attention needs to be given by owners and occupiers, prior to its implementation.

Examples of current issues at the time of publication include reform of the Landlord and Tenant Act, reform of the planning system, and progress with the Commercial Lease Code.

Case law

The general criteria for knowledge of case law is current case law, major case law, case law mentioned in written submissions, and case law underpinning key points made in the critical analysis.

As an example of how easily candidates can misdirect their preparation towards the interview is by obtaining a copy of APC course notes, such as from a university course. Numerous cases can sometimes be listed, which are excellent for reference and study, but it is not necessary to learn about all cases for the APC interview. At any one time, there might be 10–15 cases that could possibly arise in the interview. Even then, assessors are establishing that candidates are able to perform in practice, and will forgive gaps in knowledge of case law if the candidate appears sound overall. Exceptions, however, could be candidates who are unaware of a case despite its extensive reporting in professional journals (suggesting that they do not keep up to date by reading journals), and candidates referring to case law in a critical analysis which they have interpreted incorrectly, and/or seem not to understand under interview questioning.

Public sector/in-house

Public sector and other in-house candidates (such as those working for corporate property teams) sometimes comment that the APC is orientated towards private practice. Given that one of the primary roles of RICS is to uphold standards for the profession's clients, it is imperative that the process of becoming professionally qualified reflects surveyors' interface with clients.

Where a candidate has not had experience in working directly for a client, assessors will reflect this in their questioning, and in their decision to pass or refer candidates. However, it is important that in-house surveyors show an appreciation of the relevant issues, in the same way that all candidates have some gaps in experience, and seek redress through professional development.

As an illustration of how in conjunction with appropriate training support APC success can be achieved, the structured training programme implemented by Birmingham Property Services between Spring 1999 and Spring 2003 secured a 13 out of 13 pass rate. Jacky Gutteridge, BPS's training and development manager complemented day-to-day guidance from managers with external training support. As well as covering the usual subjects, emphasis was given to the issues affecting property investors/landlords, commercial property occupiers and private practice consultants in addition to in-house surveyors.

Chapter 11

Examples of Interview Questions and Answers

An important factor to be aware of if using examples of interview questions when preparing for the actual interview, is that they should be considered alongside the real study of the subject-matter.

If revision, on preparation for the interview, includes learning answers which could be given to certain questions, then it will be evident to assessors that candidates do not have the required experience and knowledge.

As mentioned in previous chapters, interview questioning relates to candidates' areas of experience and competencies – although a knowledge of wider issues is still important, including professional ethics and current industry issues.

It is difficult to illustrate what constitutes a level 2 or a level 3 question. This is because the judgment by assessors as to whether a candidate is at level 2 or level 3 is really dependent on the quality of their response.

This chapter begins with extracts from previous *Estates Gazette* 'Mainly for Students' articles, and then provides further examples of interview questions, and in some cases, guidance on answers. Some of the questions are supplied by Midlands Property Training Centre who work closely with candidates and assessors on their experience of the APC interview. A series of rent review questions are then provided, based on a free of charge event facilitated by Birmingham Property Services, the in-house property consultancy of Birmingham City Council, for all APC candidates. Further examples of questions are included within MPTC's free of charge updates for general practice surveyors: candidates may request these from apc@mptcentre.org.

Mainly for Students extracts

Q Did you need planning permission for the advertising boards on the outside of the property you were letting? Why not? Under what legislation does this fall? In what circumstances might you still need planning permission even though the boards are less than the maximum size?

Q I see that the marketing particulars include the date when they were produced – why is this included? Do you consider that you are acting in your clients' best interests if you are letting prospective tenants know that the property has been on the market for a year?

Q What sort of marketing strategy would you recommend for the letting of a high street retail property expected to secure a rent of around £25,000 pa?

Q Assuming that you have been asked by a tenant to find a property (retail, office, industrial, residential – location, size possibly described), what initial steps would you take?

Q If you were examining comparable evidence for offices recently let on a 10-year lease with a rent review after five years, and, following a rent-free period of one year, the rent was to be £10,000, how would you analyse that transaction to establish a day-one/effective rent? An easy answer would be £8,000 (£10,000 × 4 years = £40,000 ÷ 5 years = £8,000). You might like to elaborate and say that a present value calculation could be done. Comment could be made on the length of time over which the incentive is attributed – to the first review, or the duration of the lease, for example. Break provisions could be relevant.

Q If market conditions were poor, and a client asked you to dispose of an investment property situated in a secondary location, what factors would you consider in recommending an appropriate method of disposal?

Q You mentioned in your critical analysis that it was felt better to expose the property to the market rather than undertake the rent review and then sell the property. What factors influenced your decision?

Q If you found suitable office accommodation on behalf of a client who was particularly concerned about potential service charge liabilities, what investigations would you undertake in order to establish whether the service charge arrangements did not present undue risk?

Q With reference to your search for suitable office accommodation on behalf of a prospective tenant, you encountered problems in respect of consent being required from the superior landlord. There seemed to be a number of similarly suitable properties on the market that did not require landlord's consent. On reflection, what do you feel that you have learnt from this? (The question is slightly critical of the candidate's possible lack of aptitude in respect of the difficulties that can be caused when superior landlord's consent is required. Candidates should understand: the effect of lease terms; possible difficulties in securing consent; the reasons landlords include protective provisions in the lease (upholding investment value, management reasons); how to deal with an unreasonable landlord; the precise scope of the alienation provisions, and what may or may not constitute consent being unreasonably withheld. The practical issues associated with the need for superior landlord's consent also need to be addressed, such as delay and risk of losing the tenant – especially when there could be similar properties that involve negotiating with the landlord – lease terms being more negotiable than if taking a sublease.)

Q What is the effect of an 'absolute' user clause? Give me an example of a 'qualified' user clause. If a tenant seeks your advice in respect of the effect of a 'fully qualified' user clause on rental value, what do you tell him, bearing in mind that he is struggling to run a butcher's shop and the landlord is presenting evidence of rents achieved for more valuable retail uses? (Candidates specialising in sales and lettings/agency work are clearly required to demonstrate more than sales and marketing skills.)

Q What might be the purpose of a valuation adopting a special assumption of vacant possession, and who might require such a valuation, and why?

Q Are there any situations you can think of where you are assessing value but are not required to produce a valuation in accordance with the Red Book? (These are basic questions that candidates should be able to answer. They demonstrate how a relatively simple question requires a good-quality answer. They also demonstrate how questioning can lead to more difficult questions that test a candidate's understanding, as opposed to an ability to repeat standard definitions. Again, questions could relate specifically to work outlined in the summary of progress or the critical analysis. If, for example, you have disposed of a contaminated site, you could be required to comment on the issues affecting the length of the restricted marketing period – such as prospective purchasers requiring sufficient time to establish the extent of contamination and its effect on value, and developers discounting disproportionately highly for any uncertainty. While on the topic of contamination, an assessor could, for example, ask, 'What obligations are placed on owners of contaminated land in respect of current legislation?', 'Can you outline the current legislative provisions in relation to contaminated sites?' or, 'You mentioned in your submissions that you commissioned a contamination survey in respect of a potential residential development site. What did the survey find?')

Q How would you value an office building that has a rent passing approximately 20% above market value, having 22 years to lease expiry, with three years to the next review, and let to a blue-chip covenant? (The reference to blue-chip covenant helps you out. The valuation of a prime office building let to a blue-chip covenant would be different from the valuation of a poor-quality office building let to a tenant in financial difficulty – especially as the property is overrented. However, where an assessor does not give a clue as to the type of building, do not jump to the conclusion that it is a prime investment situation. Briefly outline the considerations and state, for example, 'initially I would establish factors such as the location, quality of the property and the covenant strength of the tenant, but assuming this was a prime property, I would value the interest by ...'. Generally, there is a limit to the back-tracking that would be undertaken. In the above example, you would, in practice, establish the other lease terms, but if this has been covered already in the interview, it may be unnecessary to mention it in answers thereafter. Similarly, for example, it would not really be necessary to repeatedly begin an answer by mentioning the need to agree terms of instruction in writing, if this has already been covered.)

Q If a new client requested that you banked a substantial amount of cash for him in your account, what suspicions should this raise, and what action might you be required to take?

Q If you are acting for a landlord of a modern office building (having no redevelopment potential or other reason for the landlord to be expected to wish to secure possession) in respect of a rent review, and the tenant's agent claims that the rent should be heavily discounted because there is a short term of three years to lease expiry, what arguments would you present back to the tenant's representative?

Q With reference to the restrictive covenants you referred to with respect to the potential B1 development opportunity, what are the general grounds for being able to discharge or modify restrictive covenants?

Q How might a condition of disrepair affect the rent payable at rent review, first, assuming that the tenant is responsible for repairs and, second, when the landlord is responsible?

Q If a client were to take on a lease of office space within a 1960s building, what enquiries would you be making to ensure that your client was not exposed to a significant liability for service charges?

Q In what circumstances might a landlord wish to include a 'keep open' covenant in the lease?

Q What planning considerations could apply to a development opportunity for a warehousing/distribution depot situated in a relatively built-up area?

Q What are PPG 6 and PPG 13 all about?

Q Briefly outline the opportunities which might be available for an occupier to minimise their liability for rates.

Q What does *rebus sic stantibus* mean?

Q It was mentioned in one of your submissions that you inspected a number of residential properties with a senior colleague. How might you detect signs of rising damp? What might cause it? What could cause condensation? How could you distinguish between rising damp and condensation? What might be the cause of rainwater penetration? What is the difference between wet rot and dry rot? Give an example of how problems may occur with a flat roof.

Q What could cause cracking to brickwork?

Q If a plot of land is being acquired by CPO powers which does not have the benefit of any planning consent, but it seems that you may be able to build a house on the plot, how would you attempt to provide evidence that consent could be secured and that development value should be paid?

Q Briefly outline the 'six valuation rules' for compulsory purchase.

Q What was the case *Stokes* v *Cambridge* about?

Clearly a vast range of questions could be asked at interview. Questions could range from a detailed examination of a candidate's knowledge that supports work specifically referred to in a report submission, to general questions covering areas that a candidate may not have had practical experience of, but, as a chartered surveyor, should have some knowledge of. This should be borne in mind when considering how difficult or relevant some of the above questions may be.

Q With reference to your presentation, upon what basis was the property measured? What areas or items were excluded?

Q What rates per m²/sq ft are 1,000m²/10,000 sq ft industrial units fetching? What about office rents? What sort of incentives are tenants able to command? What length of lease is considered to be the market norm? Are there any developments planned in the area that are likely to affect values over the next few years?

Q Did you investigate the possibility of changing the use of the ground-floor offices to a more valuable retail/leisure/restaurant use given the development of such uses in neighbouring properties?

Q You said that you acted on behalf of a liquidator. What are the differences between a liquidator and a receiver?

Q You sought legal opinion on the obligation on your client to use 'best endeavours'. What was the meaning of the term and how might it differ from 'reasonable endeavours'?

Q When you assisted with the computerisation of certain office records, what regard did you give to the Data Protection Act?

Miscellaneous examples

This section includes various examples of interview questions and, in some cases guidance on answers.

The initial illustration relates to a candidate being asked about their property marketing work by an assessor. 'A' is assessor and 'C' is candidate.

A   You commented in your summary of progress for the second year that you had undertaken a letting of a shop in Birmingham city centre. Can you first please explain a little more about the location and the size of the shop?

(Here the assessor requires more information from the candidate before being able to judge the quality of their answers, and their understanding of their work in practice.)

C   The shop was situated on Corporation Street, which is located off New Street, the main shopping area in Birmingham – although there are other major developments such as the new Bull Ring. It measured approximately 70 m²/ 750 sq ft.

(As an example of how a candidate could elevate their answer, and further sell their knowledge to assessors, but without talking at undue length on issues unrelated to the question being asked, the candidate may continue as below.)

Surrounding properties tend not to be occupied by the national retailers, although there are still some national covenants, which influenced the national marketing campaign recommended to the client. Because they were looking to sell within the next year or two as part of retirement planning, it was important that in addition to securing a letting at the right rent and lease terms, that the investment/capital value was upheld.

(The candidate has taken the opportunity to sell to assessors their recognition of the importance of assessing client's objectives, and of the link between lettings work and investment value.)

A   Assuming then that you have taken instructions etc. and you are inspecting the property, what were some of the key things you were looking for?

(Note how such a question seeks to establish the candidate's applied knowledge. A similar question could have been a scenario, such as an assessor referring to a secondary retail property outside a main retailing area. The candidate who delivers a pre-rehearsed list of tasks they would undertake on inspecting a property, and which includes aspects not relevant to retail, and misses key aspects associated with retail property, would fail to give an answer to the required standard. This is also an example of where a candidate can feel that they have given a good answer because of the amount of information they have provided, but the information does not relate to the question being asked, and also, there may be many issues beyond those assembled by the candidate as part of their preparation for the interview.)

C   It would have been factors which determined the types of tenants to which marketing may be targeted, factors actually affecting marketability and factors affecting value. It was also important to establish the potential for any alternative uses which would be more valuable than the existing use – such as the possibility of securing A3 consent instead of A1, although two previous planning refusals indicated this would not be achievable. Location and surrounding uses were important, as was a broad indication of footfall, such as that generated by the location of public transport or major department stores, for example. A note was taken of other vacant retail premises in the area. The property was measured on a net internal basis, and zoning calculations later undertaken. The accommodation comprised a basic retail area which although not a shell, would still require decorating by an incoming tenant, with a suitable floor covering also to be added. Kitchen and storage facilities were in good condition, although storage space was limited. Access and loading difficulties were evident from the regularly congested entrance at the rear. The property provided a reasonable width to depth ratio, and a good glazed area to maximise prominence. There was also scope to replace the current signage, and ensure prominence. I did in fact check that the property was not a listed building or in a conservation area, which was thought necessary because of its age and the character of nearby properties, but this was not a problem. This also allowed an advertising board to be erected. From the office, I checked other issues such as the rateable value of the property, and whether the landlord's interest was actually freehold rather than long leasehold, as this could influence the use of the property or the scope to undertake alterations.

(As well as covering good points, the answer has been elevated with the final sentence. Because the question was about the factors considered on site, this additional comment is kept brief. Also, candidates will not be able to cover every point in their response, and other points could no doubt be added to the above.)

A   You mentioned A1 and A3. Can you please provide more detail as to what uses these are, and also what the planning issues might be in securing A3?

(This is an example of how comments by a candidate can lead to follow up questions.)

The assessor could continue questioning in areas such as the marketing campaign recommended, the marketing budget, the cost of individual advertising outlets, rental valuation, zoning methodology, which areas were included and excluded from the measurement, whether any particular valuation adjustments were made, how might the valuation be approached if there were offices above which were part of the demise, and how covenant strength was measured.

In now drawing on some of the material provided by Midlands Property Training Centre, further examples of interview questioning are set out below.

A   Thank you (following presentation). Just remind me what the area of the property was.

C   1,500 sq ft, which is about 140 m².

(Areas are now quoted in square metres, although reference to square feet often continues. Assessors are likely to be more familiar with square feet, but will still require you to understand the requirements in respect of measurements stated in marketing details, for example.)

(The critical analysis assumed here is a rent review of shop premises in the high street of a large town, undertaken by the candidate on behalf of the tenant.)

(The example is of a retail property, but candidates should understand the following issues in respect of other property types – especially those with which they have been involved.)

A How did you measure the property?

C In accordance with the RICS Code of Measuring Practice which, for retail property, is a net internal basis.

A What areas did you include and exclude?

C Items excluded from the NIA for the shop were a toilet, a cupboard containing service facilities, an internal structural wall, the stairs to the first floor and a storage area because its headroom was below 1.5 m high throughout. Items included in the measurement were a kitchen (because it occupied, and was accessed from useable space), and the area occupied by internal partitions erected by the tenant.

A Where did you run your tape measure to within the unit?

C I measured the shop from wall to wall – in other words measuring the shell of the property. It is important to ensure that the area taken by tenant's fittings is included in the measurement.

A Why do we have to measure properties accurately?

C One reason is that the measurement of the property determines its valuation. As well as the area being accurate, the valuation is affected by the approach taken, for example, to zoning, or to first and other floors – and also the uses to which they can be put. Another reason is the Property Misdescriptions Act 1991, which requires advertised areas to be accurate. If applicants are mislead, the Office of Fair Trading may take action against an agent. This may, for example, arise where on inspecting a property, applicants are disappointed at the difference between the advertised measurements, and the actual measurements. When taking on a case, I would usually satisfy myself that the measurements are correct, and not, for example, rely on file measurements if it was a property already in our management. However, for the larger interests, it would not be cost-effective to re-measure the property, and I would rely on the firm's measurements previously undertaken – although I would, of course, check whether there had been any alterations, additions, improvements, etc.

(The candidate is providing general comment, but showing assessors the common sense, professional approach taken to their work. The candidate is on the verge of waffling though, and should draw the answer to a close here.)

A You mentioned zoning. Tell me a bit more about this.

C Shops are usually measured for valuation purposes in six metre zones, although the method depends on local practice (rather than the Code of Measuring Practice specifying the methodology). Some small shops, particularly those in poorer quality locations, may be measured on a flat rate basis.

A I saw from your summary of progress that you had helped a colleague with the measurement of a department store. Did you zone that?

C No, that was done on an overall rate.

(The above examples relate to a shop, but candidates must similarly be familiar with the measurement of all of the properties with which they have had dealings – and also all main property types – industrial, offices and retail.

The detailed answer in respect of measurement demonstrates how candidates can equip themselves to deal with questions at the APC interview. Some candidates attend the interview ill-prepared, and will struggle to provide an answer of the required standard in respect of measurement – sometimes because they are unaware of the precise basis, and have just drifted through their case work. You should ensure that you able to provide an answer to a question such as such, 'What areas did you include and exclude' and to the above question beginning, 'You mentioned zoning'.)

The following examples move on to rental valuation and comparable evidence, and then rent review and third party issues.

A What are market rental values in the high street for shops the size of the subject property?

C About £25 per sq ft/£270 per m² on a zone A basis. This reflects comparable rental evidence, and also the general views of agents. There was only one letting that provided evidence, but that derived from a lease surrender, and a re-grant to a new tenant, raising questions about how representative the deal was of market value, especially as the parties' negotiating positions may have been influenced by rights under the existing lease. A couple of rent review transactions were more helpful.

A Yes. Yet I see from your appendix in the critical analysis that you have detailed five transactions. How would you rank the quality of evidence generally?

C Open market lettings would generally be the best guide as to what a tenant is prepared to pay. Rent reviews would be the next best, and then lease renewals. Rent reviews are relatively straightforward, but lease renewals can be affected by statutory rights, the ability for the tenant to vacate and the scope for tactical manoeuvring. As mentioned, I would take care with surrenders and re-grants, and also, in some cases, with contracted out tenancies and short term lettings.

A One of your comparables was a short term letting, and another was a contracted out tenancy. What was the problem with the short term letting?

C The landlord had granted a short term tenancy to a 'pound per item' type firm who occupy premises for short periods, at concessionary rents, on the basis that they vacate as soon as the landlord requires possession. The landlord receives an income, and the property is kept secure by a tenant who has established a reputation in the region for ensuring that the property is left in good condition. Such a short period did not compare with the assumptions relevant to the subject rent review – i.e. that a term of 15 years was to be assumed (because this was the unexpired term of the lease).

A What about the contracted out letting – and also what does 'contracted out' mean exactly?

C First, this letting was not on the high street, and an adjustment would also have been required in respect of an inferior location, enjoying less trade. The landlord was a local investor who typically attracts tenants who do not seek professional representation. He grants contracted out tenancies in order to avoid granting security of tenure on a building which may have conversion potential to flats one day (but only if he can acquire the light industrial unit and storage yard at the rear). He also grants contracted out tenancies because of the favourable position this can present at lease expiry (where the tenant has no security of tenure). A higher rent, and more favourable other terms, can be secured than if the statutory position applied – or indeed than if the property was exposed to the market in the normal way – as per the hypothetical rent review provisions in the lease, for example. Contracting out refers to the exclusion of sections 24–28 of the Landlord and Tenant Act 1954.

A If the definition of market value is the amount that a tenant is prepared to pay, and the tenant of the property to which you refer is prepared to pay, say £30 per sq ft for a new contracted out tenancy, why isn't this market value?

C Because the basis of market rental value for rent review, for example, is that the property is assumed to be vacant and to let, with the current tenant's occupation being disregarded. Also disregarded is any goodwill generated by the tenant. A tenant's willingness to pay £30 per sq ft for a new tenancy, as against having to relocate (at expense and uncertainty), or close the business, reflects the tenant's occupation and goodwill.

A What were the main negotiating issues then?

C The effective date of the subject review was four months after that of the comparable property, and the landlord's agent felt that an uplift should be reflected. I resisted this, requesting that the landlord's agent provided evidence. Another difference was that the subject property had a five-year rent review pattern, whereas the comparable was of only three years. The landlord's agent maintained that a five-year period was more valuable to a tenant than a three-year period, particularly in market conditions where values are rising strongly. I pointed out that inflation was currently low, economic conditions relatively stable, and that although there had been rental growth over the last five or so years, this largely reflected the market's emergence from recession, and also that rental growth in retail rents was not expected to be substantial over the next few years – as indicated in the research of a large national practice. Another thing I checked initially was whether the rent could go up or down at review. For both leases it was upwards only. I held my ground, and agreed a rent of £25 per sq ft zone A.

A What did the lease say in respect of third party determination?

C If the parties could not agree the rent within one month after the tenant had served counter notice to the landlord's original rent notice, either party could seek the appointment of an arbitrator.

A What are the main differences between an arbitrator and an independent expert then, in terms of their dealing with your rent review?

C The arbitrator considers the evidence presented by the parties, whereas the independent expert determines the revised rent from their knowledge and expertise (although may invite evidence from the parties). If, for example, there is no evidence, it may be difficult to secure an increase at review with an arbitrator, whereas the expert can use his judgment, and possibly also his knowledge of previous market transactions in order to determine the rent.

A What if you do not like the arbitrator's decision?

C You cannot appeal against the opinion of value, but could appeal on points of law which you felt had been determined incorrectly by the arbitrator.

(The assessor may check that the candidate knows exactly what is meant by a point of law, and give examples, as opposed to simply picking up the phrase from revision notes.)

The following questions concentrate on rating.

A You mentioned in your marketing particulars included in the appendix of the critical analysis that the rates payable were £2,356 for the year. How did you arrive at this figure? (Basics: Rateable value × UBR = rates payable. All candidates should know this.)

A What is the Uniform Business rate and how is it established?

A What is the UBR for next year? (If a March interview.)

A We have talked about the UBR, but what else influences how we derive the actual amount of money a client/property occupier pays in rates?
   (For example, transition, empty rates, range of appeals/proposals. This is a good example where the level a candidate is at, i.e. 2 or 3, is judged by the quality of the response, rather than the question being to level 2 or 3. This is a relatively straightforward question, and something as simple may not be put to a candidate with level 3 in rating – although it may be a gentle opener, bearing in mind that assessors may open questioning on a new subject with an easy question to enable discussions to commence.)

A When was the Antecedent Valuation Date for the 2000 revaluation?
   (This is a relatively straightforward question – really for a candidate without level 2 or 3 as a rating competency – but note below how the questioning develops. Note: Some of the questions will depend on the timing of the interview, and this document may be used by candidates several years after publication.)

A Which property types, and locations, have you found to have increased quite considerably in value between April 1993 and 1998 – and by approximately how much?

A What sectors/locations have seen strong growth between 1998 and 2003?

A I see from your summary of progress that you have dealt mainly with offices in Leeds city centre. What increased liability, if any, are your clients likely to face/have you advised them on approximate movements in value in order that they draw such information into their operational budgeting?

A Assuming that you have been asked by a client if you could act on their behalf in respect of an appeal against their RV, can you run through the process?
   (This is a good example of a question that is straightforward to answer if good experience has been gained, as the candidate can picture themselves going through the process in practice. It is harder to recall processes if limited experience has been gained, and there is reliance on notes.)

A So what is the definition/assumptions of rateable value – Rating (Valuation) Act 1999, amending Schedule 6 of the Local Government and Finance Act 1988?

A If you were instructed to minimise the rates liability of an old warehouse, built in the 1930s, and used for the storage of scrap cars, with leaks in the roof and the general condition of the property not being relevant to the typical occupier, what arguments would you present to the VOA to minimise rates?

A If you cannot settle with the VOA, who do you appeal to?

A Then where?

A Can you appeal to the Court of Appeal and/or House of Lords on all points?

A Can you give me an example of a point of law?

A Even though it may be cost-prohibitive to go to an appeal in relation to an individual case, there may still be benefits in doing so. What am I thinking of?

A When you appear at a Valuation Tribunal, what exactly is your role?

A What are the implications of this/can you fight for your client's best financial position?

A Could there be any implications if you are on an incentivised fee basis?

A You have also got the Landlord and Tenant competency to level 3, and also the arbitration/dispute resolution competency to level 2 (both of which include rent review) – you also have to take an expert witness role there – how have you seen your managers approach this/how have you drafted reports on their behalf?

   (There are two points here: the assessor making the links between the candidates' competencies, and the assessor acknowledging that the candidate is unlikely to have had an active personal involvement in an arbitration – but nevertheless expects the candidate to have a reasonable working knowledge through having observed colleagues.)

A What is meant by year to year assumption?

A Does this infer that security of tenure is limited?

A Do you make allowance between, say a rent review frequency of five years in a comparable, and the assumption, for rateable value purposes, of one year?

A If not, would you make an adjustment if the comparable was 14-yearly reviews?

A It was noticed from your summary of progress that you acted on behalf of a national building contractor in respect of a number of sites where portacabins were present. What is the general test of whether rates should be payable/was there any case law that you drew on?

The following questions cover estate management.

A What type of items are included in service charges of office properties – say the ones referred to in your summary of progress?

C Communal expenses, such as water, drainage, electricity, gas, rates, insurance (which may be separately chargeable). Cleaning and maintenance of plant, air conditioning, lifts, the exterior of the building, car parks, roads etc. Management fees. Receptionists, security staff, etc. Landscaping, plants and other attractions.

A What type of items would the landlord usually be responsible for?

C Tenants would not really be responsible for capital items, such as those relating to the original development of the property, or expenditure which is in excess of maintenance and repair (although replacement may sometimes be the most cost-effective remedy). The extent of a tenant's responsibilities could depend on the type of property, and typical terms of occupation. Tenants of larger offices in city centres, holding 15–25 year leases, are likely to have more items covered in the service charge than tenants in smaller, secondary, offices, who may stay for shorter periods. They do not wish to be exposed to upwards service charge variations.

A What is a 'sinking fund'?

C A sinking fund provides for infrequent, and usually large, items of expenditure, and/or for emergencies. This helps consistent service charge levels to be imposed from year to year – as opposed to tenants facing substantial increases when major repairs are required.

A What about 'sweeping up provisions'?

C It is means of incorporating elements in the service charge which are not expressly provided in the lease, but which reflect the overall nature of the services to be provided.

A Briefly, what would you consider in respect of service charges if negotiating the leasing of office space on behalf of a client?

C Establish the current level of service charges. Consider service charge levels payable for similar properties and consider rent and service charges as a total overhead (with high service charges, for example, effectively lowering rental values). Obtain the service charge accounts over previous years. Find out the basis of apportionment to the subject interest. Inspect the building to establish how effective the management of service charges appears to be. Find out the level of sinking fund. Establish whether this is adequate, and judge whether there are likely to be significant increases in service charges over the coming years.

A If you are acting for a tenant of a 25-year lease with five years to expiry, who receives a notice from a landlord stating that certain works must take place to the property, and if not, the lease will be forfeited, what would your advice be to the tenant?

C I would advise them that a landlord wishing to forfeit the lease for disrepair would have to serve a 'section 146' notice on the tenant. This does not have to be in a prescribed form, but should say why the tenant is considered to be in breach of the repair covenant, and outline what has to be done in order to remedy the breach. The tenant must be given a reasonable period of time to undertake the works. In the example described, the Leasehold Property Repairs Act 1938 would be relevant, as the lease was granted for a term of seven years or more, and there are at least three years to expiry. The Act requires landlords to obtain permission from the court before proceeding. I would establish whether the landlord had complied with the section 146 requirements, consider the liability of the tenant having regard to the terms of the lease and the condition found on inspection, and look to open discussions with the landlord.

A What terms of the lease might help you?

C The repair provisions, which could, for example, be 'to keep the property in good and substantial repair', or, 'to keep the property in no worse condition than that at the commencement of the lease, as evidenced by the schedule of condition appended to the lease'. This would depend, of course, on the type of property.

A If the landlord seeks permission from the court, on what grounds might he apply?

C There are five grounds. 1. That the value of the landlord's interest has been damaged substantially by the state of disrepair, or if the works are not carried out immediately, it will be. 2. That repairs need to be undertaken to comply with legislation. 3. Repairs need to be undertaken to protect another occupier within the building. 4. The cost of carrying out repairs now is low compared to the cost that would be incurred if the condition deteriorated and repairs were undertaken in the future. 5. Because of special circumstances, it is just and equitable to grant leave.

A What do you understand by the term 'self-help'?

C It may be possible for the landlord to re-enter the property, carry out the works himself and charge the debt to the tenant. The classification of this amount as debt rather than damages is the critical distinction. This was the subject of a case, *Jervis* v *Harris*, in recent years.

A What would the basis of damages be for disrepair?

C The diminution in the value of the property as a result of the breach of the repair covenant, or the cost of undertaking remedial works, whichever is the lowest.

The following questions cover planning and development appraisal.

A Let's assume that you are seeking planning consent for part of your client's builders yard to be developed for residential use, and this involves the acquisition by yourself of further adjoining land from the local authority (currently being negotiated). You are seeking outline planning consent. Run me through the elements that you will complete on the forms, the requirements/annotations in respect of plans and who you may also notify?

A Can you give examples of reserved matters?

A What if the site was to draw on an attractive canal/river feature – any issues there?

A Rights of way across fields/development sites – how are these prescribed, and how might you extinguish or divert such rights in order to optimise the value of the site?

A Can you summarise issues in respect of rights of light, and how this might affect the form of development and its design?

A What aspects of title would you typically examine for a development site?

A What rules cover TPOs, and if there was not a current TPO, would you be able to chop the tree down before anyone was alerted to the possibility of preservation?

A Why might you be required to undertake a development appraisal?
   (Candidates need to be aware of the circumstances in which development appraisals are undertaken. Examples include acting on behalf of a developer to establish if a site should be acquired/how much should be paid; acting on behalf of a developer to establish whether construction should commence where the developer already owns a site; undertaking sales of development opportunities; appraising a developer's proposals on behalf of a lender; appraising a developer's proposals on behalf of regeneration agency.)

A What is the purpose of a contingency, and what type of factors would it cover for greenfield residential development?

A   What if it was a brownfield site – what factors might be relevant there?

A   So what sort of percentage contingency might be suitable for a greenfield site, and a brownfield site?

A   What might be the difference in finance cost between a development appraisal of a block of 40 flats, as against a development of 40 individual houses?

A   How would you treat obligations in respect of social housing in a development appraisal (there may be various ways)?

A   Developers' profit – what sort of percentage might you use for the development valuation?

A   Percentage of what exactly?

A   (Assuming above response was a percentage of costs) What about the approach of basing developers profit on GDV – why might that be helpful for developers?

A   At the end of the valuation we have run through, there will be the residual land value element. Is this the end of things? (Purchasers' costs – stamp duty, surveyor's fees, legal fees.)

A   Moving onto aspects related to commercial development appraisals – what might be the benefits of a pre-let – say to an office development scheme in a city centre?

A   Why might it be best to secure anchor tenants – let's say in an office development scheme, and then leave other lettings for now?

A   Still looking at pre-lets, where might a pre-let (assuming of the whole) change the inputs to a development appraisal?

A   Why might the cost of finance change because of a pre-let?

A   In your commercial development appraisals, how exactly is the cost of finance calculated?

A   Which computer packages do you use for development appraisal/ development valuation?

A   In a commercial development appraisal, where do the main sensitivities lie?

A   You referred in your summary of progress to making representation to the local plan for a particular case – can you explain the issues?

A   Special Planning Guidance – who prepares this, and what weight does it carry?

A   You have just been instructed by the local authority to prepare a development brief for a five acre, potential mixed use site on the edge of the main city centre commercial/retailing area. Run through the process you would undertake?

A   Let's suppose that a client has asked you to investigate the potential for securing a residential development consent on three acres of land on the edge of a large town – population say 50,000. (a) What planning issues would be relevant, and (b) What factors would you look for on site?

   (The question is an example of a specific type of property/development that brings through certain issues. Candidates need to be similarly aware of the planning issues affecting other types of development.)

A   Explain PPG 3 and how this would be relevant?

A   How might you draw on PPG 13?

A   You referred to brownfield development – how might you still sway the planners to greenfield development even though there is brownfield space available in the town centre?

A Social housing was also something you referred to – what rules govern this, and how much social housing might be needed?

A Is social housing going to be appropriate on the edge of town, mixed with four to five bedroom executive houses?

A The developer will clearly make less money having to build social housing instead of four/five bed housing for some of the development, but might be concerned that the presence of social housing dragged down the demand for, and value of, the four/five bed houses – are there any ways that this could be overcome?

A What planning conditions could be imposed on the type of three-acre residential development we are discussing?

A Can you give an example of a planning condition that would be unreasonable for the local authority to impose?

A How is the regeneration sector structured in England?

A Which government department is responsible for regeneration policy?

A How does Europe influence the UK regeneration sector?
   (The above are basic questions – candidates must understand the structure of the sector in which they work – i.e. appreciate factors beyond the work of their individual employer.)

A You work for xxxx Regional Development Agency – what are the stated aims, objectives, etc., and what are the key points of the corporate plan?

The following questions are in respect of compulsory purchase.

A What are the six rules of valuation for CPO compensation?

C 1. The fact that the acquisition is compulsory is ignored. 2. Open market value is the basis of compensation. 3. Account is not taken of the special suitability of the property to the acquiring authority. 4. Compensation cannot include value relating to unlawful uses. 5. Equivalent reinstatement may apply where there is not a market for a property. 6. Disturbance may be payable.

A If a new road involves a builders yard being acquired, what heads of compensation may apply, and is there anything particular you would wish to consider?

C The owner, say, an owner occupier rather than an investor, would be entitled to the existing use value of the property. They would receive disturbance compensation if having to relocate, but if the business could not be relocated, compensation would be paid for the 'total extinguishment' of the business. I would establish whether the builders yard had any redevelopment potential, in the absence of the scheme, which may produce a higher value than the existing use. It would be necessary to establish whether there were any planning consents in place. If not, other planning assumptions could relate to the zoning in the local plan, and also to any general policies it contained that might support development on the subject site. It may be necessary to obtain a certificate of alternative use from the planning authority. If development value was paid, disturbance compensation would not be available, as the business would, of course, be extinguished anyway by the development. Legal fees and surveyors' costs would also be claimed.

A What if half of the builders yard was acquired, and your site could now accommodate a fast food operation as a result of the presence of the new road?

C The value of the retained land as a fast food development opportunity would be likely to substantially outweigh the value of the lost land as a builders yard. The principle of 'betterment' would mean that compensation for land value would not be payable, although the owner could require the acquiring authority to pay for accommodation works, such as a new boundary fence.

A Can you give me an example of where severance and injurious affection might apply?

C If a farmer had a field split by a new road. Both parts would be useable, but less efficiently – with the value lost as a result of this, being severance compensation. Injurious affection would account for the affects of the road, such as noise, vibration, and the discharge of solids or liquids on to the retained land.

The following are examples of questions in respect of professional ethics.

A In your critical analysis, you referred to a conflict of interest check being undertaken. What was the purpose of this?

   (This is a good example of where candidates need to be familiar with what they have stated in their written submissions. The question is, in any event, a straightforward question requiring the candidate to demonstrate a fundamental understanding of conflicts of interest. Candidates can obviously not act in practice if there is a conflict of interest.)

A What process do you go through to establish any conflicts?

A Your software systems (likely to be applicable to only large practices) – what sort of things will they not pick up which may still constitute a conflict?

A Why might it be inappropriate for a firm to manage a multi-let property in which it is a tenant?

A (Version 1) – You mentioned that you had been responsible for administering service charges in connection with the collection of rents and service charges on behalf of a client. What exactly does the RICS Rules of Conduct say about the rules here?

   (Version 2) – Why are clients' accounts necessary?

   (The first question illustrates how an interview question can be more detailed, and the expectations placed on candidates much higher, where they have undertaken a specific aspect of work.)

A What happens if someone ends their business, and retires. Does this end liability for claims from former clients, and the need to continue insurance?

A What would you do if asked to undertake a valuation of a tin mine in Cornwall?

   (The poor candidate, after the interview, complains to his friends that he/she knows nothing about tin mines or Cornwall and that the question was unfair. Candidates are expected to be alert to the fact that if they have not had experience of a specialist area of work, that they should not take on instructions. An answer could be on the lines of, 'As I have not had experience of that, I would decline the instruction. It may be possible to find someone within my firm to undertake the job, who did know about this specialist area. Alternatively, I could advise the client to contact RICS for a list of people who may be able to assist; could sub-instruct or could introduce business elsewhere.)

Birmingham Property Services APC event

The following illustration of rent review questions is based on an APC event facilitated by Birmingham Property Services on a free of charge basis for all general practice candidates in January 2003.

The event considered the various issues arising at rent review in relation to the following case example:

> The scenario is based at Alpha Tower, Birmingham – a 25-storey office block situated at the bottom of Broad Street, approximately half way between Birmingham's city centre office area, and Brindleyplace.
> A 20-year lease commencing in June 1998 is due for review with effect from 24 June 2003.
> The demise is two floors (fourth and fifth) having an area of approx. 5,000 sq ft each – total 10,000 sq ft (it is acceptable to work in sq ft rather than m².)
> Rent reviews are five-yearly.
> It appears that the office market has strengthened over recent years, and the landlord is looking forward to a substantial increase.

Questions arising were as follows:

Initial contact from client
Q In your capacity as a surveyor with a private practice, you have received a call from the landlord, asking you to act on their behalf in respect of the rent review. What are the initial issues arising?

Local authority/in-house/client-side issues
Q If you were acting on behalf of your employer on an in-house basis, appointing a private practice, what might some of the issues be?

Preliminary tasks
Q You are instructed in your private practice capacity – what are some of the preliminary tasks required? (Some may have in fact been undertaken pursuant to taking on the instruction, but will be considered here.)

Importance of background research
Q As a general point, give examples of why sufficient background research is required before an inspection of the property takes place for rent review purposes.

Estate management/retained role
Q Some clients might have required a report around 9–12 months in advance of
the effective date of the review. Why might this be?

Factors affecting rental value – property inspection
Q What might some of the factors affecting rental value be for the property (noting
that some of these may ultimately become key negotiating points)? What might
the surveyor have looked for when on site? (Lease terms are considered later.)

Wider benefits for the landlord
Q In addition to any rent increase achievable, what are the wider opportunities
for the landlord to enhance value?
Q Similarly, what tactics might be available to the landlord (this will be picked up
later).

Scrutinising the lease – planning the rent review strategy
Q What are the main terms of the lease in respect of the rent review machinery –
i.e. instigating the review, and helping determine tactics? (Valuation aspects
are considered below).
   (Increases in market values since the initial letting might suggest that there
is scope for a rent increase, but this might not necessarily be the case – why?)

Pitfalls for delaying the rent review/negotiations – tenant
Q As a general point, how can tenants/their agents get caught out by delaying
negotiations on the basis that they will have to pay the increased rent at some
point, but might as well delay it for cash flow benefits?

Factors affecting value – the lease terms
Q What are the main lease terms likely to affect rental value/comparison with
other properties/comparable evidence?

Instigating the rent review – landlord
Q How might the landlord instigate the rent review?

The tenant's response
Q How might the tenant respond?

Issues arising during negotiations
The parties will have their own comparable evidence, and issues arising might
include the following:
Q What might be the effect of vacant floors on the rent review?
Q One piece of evidence is for six floors to the same tenant – approx area 30,000
sq ft. How reliable is this for the subject rent review?
Q The tenant thinks that there is evidence of other lettings in the building, but the

landlord is unwilling to provide details (rather rely on more favourable evidence)? What can the tenant do?

Q What does the ranking/hierarchy of evidence refer to?

Q What rent review 'assumptions and disregards' may be drawn on in negotiations?

Q Why might a lease of 100,000 sq ft office contain a 'hypothetical' assumption that a minimum lease term of 15 years is always to be assumed at rent review?

Q What might the third party determination provisions state in the lease?

Red Book

Q Would the landlord and tenant have complied with the Red Book in undertaking their rental valuations?

Differences with lease renewal

Q What are some of the differences between rent review and lease renewal?

Similar valuation aspects, and negotiating aspects could apply to other areas of work, including agency (letting for landlord, acquisition for tenant), lease renewal, and rating. The investment surveyor and valuer would also scrutinise lease terms in order to establish their effect on rental and capital value.

Chapter 12

Miscellaneous

This section covers the following issues:

- Referred candidates.
- Whether to appeal if unsuccessful.
- Report writing and presentation.
- Revision/learning techniques.

Referred candidates

Referred candidates will receive a report from RICS, prepared by the chairman of the panel in conjunction with the other assessors. This will outline the reasons for referral, and sometimes provide guidance on how the necessary experience and knowledge to succeed could be gained.

RICS referral reports range from relatively positive commentaries on how close a candidate is to meeting the necessary standard (and that how, after six further months of experience and professional development/learning, success is achievable) to comments on the role of the supervisor and counsellor in allowing the candidate to sit when clearly not ready.

The referral report will outline the requirements of the candidate regarding their re-sit, such as whether a new critical analysis is required, or the original critical analysis only needs to be updated. A further 100 days of experience will usually be required, and the diary, log book and professional development record will continue to be recorded.

Templates 15, 16 and 17 will have to be completed for the re-sit (as shown in the RICS guides). Template 15 is candidates' initial commentary on how they will gain the necessary experience, and later, template 16 is how this has been achieved. Template 17 is the supervisor's and counsellor's comments on the additional experience gained. Template 15 can be completed relatively easily, but template 16 may have to be submitted to RICS for the next round of interviews only a short period after the candidate has been referred. If, for example, a candidate is referred from a late-October interview, and receives the result in mid-November, in having to submit for the next interviews in January, and allowing for the Christmas break, there tends to be little to write about in template 16. This is not however a problem, and candidates should have the chance at the next interview to demonstrate their improved experience and knowledge to assessors.

Although the referral report usually states that candidates need to gain a further 100 days of experience, RICS in practice works to 100 days before the end of the assessment period, rather than the date of the candidate's interview. This saves a London based candidate, for example, travelling to a Harrogate interview in order to allow more time.

It is important that referred candidates present themselves for the next interview as having developed their experience and knowledge from the previous interview. Unfortunately, some candidates simply undertake the same approach to revision, and gain no new experience – and consequently fail.

It is also important to remember that 45 minutes of interview questioning within a 60-minute interview can involve only a selective examination of candidates' knowledge. This is, however, sufficient to determine whether a candidate is at the necessary level – and in the very occasional situations where this is difficult, the interview may run on for another 10–15 minutes.

Questioning will cover all of the core and optional competencies, and some other areas, but will obviously not cover all of the issues potentially arising within competencies. If, for example, a general practice candidate is asked how they might value a city centre office investment property, just let at a market rent of £100,000 to the city council on a 20-year lease without break facilities – and they respond that they would take the term and reversion approach, and apply a yield of 9% to the term and 10% to the reversion – it seems clear that they do not have the necessary, fundamental understanding of investment valuation.

In the above example, the market rent should, of course, be capitalised into perpetuity, and at a yield in the region of 6–7% (assuming no unusual factors, and assuming current market conditions). The good candidate would also comment on factors such as the need to check that the rent was actually a market rent and not a headline rent following a rent-free period, and which may consequently make the property overrented.

Had it not been for the way-out yields (which show poor market awareness as well as a poor understanding of valuation) the assessors may have sought clarification from the candidate that term and reversion was actually correct, noting that for candidates undertaking detailed valuations, there could in fact be term and reversion situations they are thinking of in order to provide for yield differentials at lease expiry, and/or to build in voids. The main point here, is that assessors are unlikely to go on to other aspects of valuation if a candidate has fallen at the first hurdle.

It is important when examining the referral report not to conclude that attention needs to be given to only the deficiencies highlighted. Some candidates unfortunately do this, only to receive a different series of questions at the next interview, and a further referral report, containing different reasons for the panel's decision. It is, of course, also important that any deficiencies highlighted on previous referral reports are studied in depth for the next interview. Assessors will not go through all the points, but will usually pick up one or two. A poor impression can be created when candidates struggle on points previously highlighted.

In the preface it states:

> The national pass rate for general practice surveyors taking the final assessment interview has been around 65% over recent years (which after allowing for re-sits, equates to an approximate 50% first time pass rate). This is low, considering that one of the elements of the APC is that employers (as supervisor and counsellor) are required to declare that candidates have reached the necessary levels before putting them forward for final assessment.

Reasons for the number of unsuccessful candidates include a lack of awareness as to what exactly needs to be understood within competency areas, supervisors and counsellors signing off candidates despite not having met the required levels, and candidates being prepared to take a speculative gamble on APC success – consoling themselves that failure would still be good experience for further attempts. A related factor is candidates often focusing on two months of revision, rather than at least two years of concerted learning.

Even though candidates may have sat the interview before, they should not automatically conclude that they should sit at the next immediate opportunity. As indicated in previous chapters, the need to be ready for final assessment before sitting, cannot be stressed too strongly.

Whether to appeal if unsuccessful

It is worth considering again the key comments made at the start of Chapter 8 on the interview. This includes a key comment of, 'surveyors who are involved in graduate training, and who are also RICS APC assessors, will often comment on the disproportionate level of unfair circumstances reported by candidates, as against their own experience of sitting many interviews and observing the assessment process as a whole'.

Sometimes, despite assessors' combined experience on a three-person panel of between 50 and 100 years, as against the two years or slightly more of the candidate, disappointed candidates can sometimes be insistent that they were right and the assessors were wrong. Candidates can examine a referral report, and comment for example that, 'they were not asked a question on the Estate Agents Act', when, in fact, the assessors may have asked, 'Can you outline the relevant aspects of agency law that you would have been relevant to your work as an agent in the case?'. Candidates can also consider that they have not been asked questions in a particular competency, only for subsequent examples of questions that they did receive to confirm that the competencies were in fact covered.

Some disappointed candidates can be defensive, and attempt to preserve a certain position to colleagues and managers through embellishing points of possible contention. This can lead to appeals being submitted to RICS which lack the necessary grounds to succeed.

Candidates have a right to appeal against the decision of the panel. RICS do not formally specify grounds on which candidates may appeal, but give a broad indication that they may fall into categories of (a) administrative or procedural matters, such as the panel not being provided with the right information by RICS, (b) questioning and testing of competence concentrating too much outside the main area of training and experience, and (c) any form of discrimination.

Examples could be a panel not covering two optional competencies and the candidate not having had chance to demonstrate their overall knowledge; the panel not containing assessors with the corresponding experience to the candidate; and significant procedural or administrative faults. However, even in the very rare situations where the system may have broken down, RICS, in considering the appeal, will still give regard to wider factors. For example, if a critical analysis falls substantially short of the required standards of professional report writing, presentation and content, and/or a candidate did not meet the required level 3 in valuation, the fact that a further competency was not covered, is unlikely to influence the overall result.

What does not constitute a ground for appeal is that the candidate is thought by managers, and themselves, to be good enough to pass. The APC is about submitting written reports to the required standard, and demonstrating the required level of knowledge and understanding at the interview. Good candidates can sometimes have a bad day. There are also candidates who are very proficient in practice within the work they do, and are highly regarded by colleagues, but they must still gain the necessary levels of knowledge throughout their competency areas, and also of professional ethics and current issues.

Candidates can sometimes focus only on their interview performance when considering the merit of the assessors' decision. However, the assessors make an overall assessment, including experience and training, and the quality of written submissions.

Writing techniques and presentation

This section comprises an extract from training material supplied by Midlands Property Training Centre on behalf of training providers with whom they work. It provides only a brief reminder of key presentation and report writing issues. Candidates requiring more detailed guidance may wish to obtain literature from the business and management, or education, sections of the larger high street bookshops.

Good writing and presentation technique

The following examples of good writing techniques are perhaps rather obvious after nearly two years' work in practice, but are nevertheless summarised.

- The submissions tend to take substantially longer to prepare than originally envisaged.
- It is important that sufficient time is allowed to ensure they are of the best quality.
- Repeated proof reading is vital. Aspects requiring attention include spelling and grammar, word selection, consistency in the use of tenses, factual accuracy and overall presentation.
- However, once submerged in the reports, and having read them so many times, word-blindness means that it is very difficult to pick up all the faults.
- By reading the report out aloud to yourself, disrupted speech should help to identify poorly flowing text, and slight breathlessness to indicate unwieldy sentences.
- In addition to supervisors and counsellors, a colleague or friend could assist with proof reading.
- It is often useful for the report to be read by someone detached from candidates' case load and place of work. The need to include obvious and

essential information, such as the location of the candidate's office, can sometimes be overlooked.

- A contents page and page numbering are beneficial.
- Choose headings and sub-headings carefully.
- Do not overdo sub-numbering.
- Do not hinder the fluency of the text by unnecessary reference to appendices.
- Avoid excessively long paragraphs.
- Keep the presentation professional.
- Use an appropriate size and style of typeface – ideally 12 point Times New Roman.
- Do not make headings unduly large.
- Do not overdo the demonstration of your computer skills by the use of clever, but pointless, graphics.
- Establish suitable margins and ensure that text does not appear cramped.
- 1.5 line spacing can be attractive.

Report writing

The key elements of effective written skills are the credibility they convey to the reader, and how articulate text reinforces the points the writer is making. It is equivalent to the power of effective verbal communication.

Aside from the obvious points such as spelling, grammar, font type and size, formatting/presentation, etc., principal points include.

- Write for the reader – always consider how the reader deals with your work. They require a quick, fluent and informative read, getting out of the report what they need to know. Reports, and very often candidates' APC submissions, are written for the writer – with information being left out because it remains in the writer's mind – not thinking of the reader. Assuming knowledge, both in terms of background information, and technical areas, can also leave the reader lost very early in a report. Frustrations need to be eliminated, such as having to turn back pages under the impression that something has been missed – leading to further frustrations when information cannot be found.
- Good reports outline at the outset what is going to be said thereafter. The reader then feels settled, being aware of, and at ease with, the structure of the report.
- Reports need to build up a picture chronologically – almost telling a story that retains the captivation of the reader.
- A sharper articulate style avoids an unduly high number of words saying something that could be said in less than half the words. This is sometimes a trait of APC candidates' submissions. As an example:

On arriving at the site to undertake a site inspection, my first task was to establish that the site boundaries matched the plans provided to me by my drawing department – with whom I always liaise before undertaking a site inspection.

After checking that there were no problems with site boundaries in accordance with the plans, it was necessary for me to undertake a survey of the outside of the building.

My inspection of the building was particularly important in respect of the roof where a number of slates were missing. At certain parts of the roof, it was also evident that there were signs of water ingress. Another important aspect of the inspection was recording the detail of cracking above one of the windows. Because I had not experienced cracking in such properties previously, it was important to refer the matter to my manager who would ensure that all the necessary steps were undertaken to ensure that the cracking did not present a problem.

Note:
- Half repetition of previous sentences.
- Other repetitions.
- Over use of words including in same sentence: undertake, important, ensure. Word selection is important.
- Unnecessary padding – e.g. 'it was also evident that'.
- (Another trait of APC submissions is inconsistent tenses.)

As an illustration of improvement:

After firstly checking that the site boundaries matched the boundaries on the plan, an external survey gave particular attention to the roof, as there were a number of slates missing, and signs of water ingress. Cracking was also discovered above a window, and in view of my lack of experience in this area, it was reported to my manager.

(59 words, whereas the above was 164 – i.e. a better written version takes as third as many words).

For APC written submissions, where there is a limited word count, an effective written style creates the opportunity to cover issues in more depth – better selling the candidate to the assessors. For business reports it means a much more dynamic read, such as for a client. Fewer words also means less time is taken to prepare the report, fewer hours spent on the project, and better fee/profit rates. It does, of course, need to be ensured that the text does not become too intense, such as long, list-type sentences aiming to fit as much as possible into as few as possible words.

Note also above how the title, 'Site inspection', sets up the text which is to follow. Headings and their precise wording can therefore be used within a report to create effective structure and fluency. Another, finer means of polished report writing is to lead into a new section/heading with compatible comment, or use of a linking word. For example:

Initial tasks included establishing that there were no conflicts of interest, exchanging letters of instruction, notifying colleagues of the new client and undertaking a site inspection.

Site inspection ...

Apostrophes

Many APC candidates make mistakes with an absence of, or incorrect inclusion of, apostrophes:

Examples include the following, with the correct version appearing in brackets.

- The clients objectives (client's if one client, clients' if more than one client/plural).
- The companys office in London (company's – not companies).
- The companies offices throughout the country (companies' – as more than one/plural).
- RICS had it's Christmas party last week (its – no apostrophe).
- James bike (James' or James's but not Jame's).
- Built in the 1960's (1960s – often seen as 1960's, but plural (1960, 61, 62 etc.) not possessive).
- Dilapidation's (dilapidations).
- The solicitors friends desk (solicitor's friend's).

Revision/learning techniques

This section is extracted from training material provided by Midlands Property Training Centre.

The big issues here are:

- How to learn for an interview rather than an exam.
- Making revision time productive.
- How to think quickly.
- How to recall information under the pressure of the interview.

Reading and re-reading notes is not the way to pass the APC. Familiarity with notes does not mean that candidates understand the issues. The acid test is whether candidates can respond to questions.

Mock interviews and even simple questioning by colleagues can illustrate how difficult answering questions can be.

A much higher knowledge and understanding is required to answer verbal questions, than to respond by way of essays, as with university exams. University exams can sometimes be passed by memory rather than understanding.

A pro-active approach to revision is important – making notes, and condensing notes. The fewer the number of words required, the better is the understanding.

Reading can still take place around a subject. Notes do not need to be made of everything.

It is essential in the interview to be able to think quickly.

The mind often needs a prompt.

A bullet point approach to learning and revision therefore serves candidates well. In preparing lists of key points, candidates are well positioned to give answers.

It is very difficult to produce answers from a text book memory of subjects – experience is the key – albeit complemented by learning.

Bear in mind the pressure of the interview, and the difficulty in recalling information. The above point regarding prompts will be important.

Revision time is made productive by creating a learning/revision programme in relation to the competencies and main areas of experience.

More detailed knowledge is required in the main competencies than in other competencies, and also than in wider areas where only a insight/appreciation may be required.

Some candidates fall into the trap of trying to revise wider, non-competency, areas in as much detail as the main competency areas. They attend the interview and struggle in the main areas, and despite their better than required knowledge in other areas, do not receive questions in those areas.

Chapter 13

Beyond the APC – Lifelong Learning, and Business Skills

The APC is really only the starting point for surveyors to build a successful career. It is not the point at which learning stops.

Aside from RICS proposals in respect of a post-graduate business qualification (which is under review at the time of publication) newly qualified surveyors will be required to undertake continuing professional development (CPD), and generally embrace RICS concept of lifelong learning.

CPD is the equivalent to professional development undertaken by APC candidates as part of their training period. Qualified surveyors undertake a minimum of 60 hours CPD over any three-year period, equating to 20 hours per year. RICS, however, stresses the importance of the quality of study activities, as opposed to only the logging of hours to meet requirements. Also, since January 2004, newly qualified surveyors are required by RICS to record their CPD on-line via the RICS website.

Lifelong learning means that at every stage of their career, surveyors evaluate their personal and business development aspirations, and identify suitable learning activities that will help such aspirations to be realised.

The RICS Agenda for Change in the late 1990s identified the need for surveyors to possess greater business acumen alongside mainstream surveying skills. This is to ensure that a high quality service can be provided to the profession's clients, and also that surveyors can contribute to the profitability and efficiency of their employer's business. This applies equally to the public sector as well as to private practice, and will be especially important for those surveyors looking to establish, or are already running, their own small business.

In view of the above, for most surveyors, CPD comprises a combination of training in business and managerial subjects, and in surveying subjects. This is summarised by the following commentary from RICS lifelong learning promotional material:

> For some businesses, lifelong learning will simply involve remaining up-to-date with changes in practice and legislation in order to ensure that a relatively static client base is serviced. Indeed, among the many small surveying practices in the UK and also abroad, many undertake work in similar areas, year-on-year.
>
> For other businesses however, lifelong learning extends to strategic growth, achieved through the development of new areas of expertise. Here, lifelong learning incorporates the development of effective marketing and business development techniques, including improvements in profile which help secure better quality work, more prestigious clients and higher financial rewards. Business development can, in fact, be facilitated through training by the

delivery of seminars to prospective clients, and other surveyors, and also, for example, through contributions to professional journals.

All businesses must keep up to date with changes in economic and property market conditions and the impact this can have on their clients and their own business. Factors such as developments in technology can also affect the competitiveness and viability of existing activities, but can facilitate more profitable opportunities for those remaining most up-to-date.

Set out below, with the assistance of guest contributors, is further information on CPD activities and business skills.

CPD: qualifying activities

This section was originally published in *Estates Gazette*'s 'Mainly for Students' feature of 16 September 1999, for which Margaret Harris of RICS was joint author.

The most effective training programmes are those which reflect the nature of an organisation and the roles of individual personnel. A local authority or other in-house property team, for example, may see individual surveyors covering rent reviews, lease renewals and other landlord and tenant cases, as well as sales and lettings work, together with involvement in areas such as rating or asset valuation.

Surveyors in smaller private practices may cover a similar range of work, but surveyors with the larger firms may have relatively specialised roles – and therefore have very different training requirements.

Most organisations possess a level of expertise among their surveyors which can help develop the skills of others without the expense of formal training events. Surveyors' involvement in training could range from the day-to-day support provided for junior staff to the devising and delivery of training of qualified colleagues. Clients could be invited to talks, as could surveyors from other firms.

A company's solicitors could be encouraged to deliver seminars. Many already see the commercial benefit of doing so, and surveying practices could do more in this respect.

The organisation's provision of refreshments can be an effective way of ensuring that some training events take place outside fee-earning hours, with surrounding social activity adding interest.

Training sessions conducted on a suitable group basis can also enable team-building skills and morale to be improved. This includes support staff and younger surveyors, for example, being seen as a more valued part of the team. Indeed, the ability for support staff to deal with basic property cases, following training, can present cost and resource flexibility to an organisation.

Softer subjects, such as negotiation, report writing and coping with change, are also important, but tend to be given lesser priority than topics that are more fee related.

When work pressures tend to take priority over the arranging of training activities, it is important that the organisation designates personnel with specific responsibility for ensuring that training does take place.

Structured training

To most qualified surveyors, structured training means much more than undertaking a minimum average of 20 hours per year of continuing professional development in accordance with the RICS requirements. 'Structured' means that training activities stem from pre-identified requirements as part of an ongoing programme, as opposed to occurring on an *ad hoc* basis.

For individual surveyors, structured training involves identifying training requirements through the RICS Personal Development Planner (now part of lifelong learning and CPD on-line recording). Attention is focused on strengths and weaknesses, and on how training can aid personal development, along with business opportunities.

For organisations, structured training involves ensuring that surveyors, and also support staff, enjoy the best opportunity to develop their skills.

Private study

RICS CPD works on the basis of self-assessment. Surveyors determine the suitability of the training initiatives themselves, and record the activities undertaken, together with the number of hours devoted.

Structured private study can qualify for up to two-thirds of RICS minimum CPD requirement. This can be a particularly effective form of study, especially as surveyors have a differing knowledge and training requirements.

Private study should generally be structured around specific subjects, or be part of wider learning initiatives. Simply reading *Estates Gazette*, would not, for example, be satisfactory, but the study of particular articles therein may be.

Articles do not require a 'CPD' or 'distance learning' type of designation, or the need to obtain certification from the provider, to count as CPD. When providers state the number of hours available, this must be regarded as indicative, with the recorded number of hours being determined by surveyors themselves.

Private study could also comprise preliminary research, and/or subsequent study around a course or seminar.

Surveyors attending courses should disseminate the information to colleagues, and can also arrange discussion groups. These can consider the various issues in relation to the work in which the organisation's surveyors are usually involved.

Training events and training providers

Courses and seminars are particularly useful for specialist requirements, and also where it is necessary to be informed of new legislation, case law, market practice and so on. Many of these courses are useful because they cover many topics within a single day.

The quality, content and suitability of training events can sometimes be difficult to determine in advance of attendance, especially with a large number of training providers in the market. Most commercial property training companies are in business because of an established demand for the quality of training they offer.

CPD on-line recording

This section is extracted from RICS guidance.

Online CPD is the process of recording CPD via the main RICS website. The web-based planner has been introduced so members can plan and record their learning activities. It is hoped that the online planner will eventually replace paper-based recording systems and that the benefits of recording this way will be recognised by members – as opposed to the more traditional means of simply logging hours.

To begin recording CPD online, visit www.rics.org/careers/cpd/online_recording.shtml (current at the time of publication).

A range of general information on CPD and guidance on personal development plans can be viewed initially, and to then begin recording, simply, click on 'online CPD recording',

To log on you will need to enter your seven digit RICS membership number, date of birth in the dd/mm/yyyy format and click on 'log me in'. You can set a password once you are in the system. If you opt for the default password, click on 'log me in'.

To fully activate being online, an entry could be registered under the unplanned learning field (add learning activity). This could be an 'induction to RICS online CPD' – 0.5 hour, based on the introductory understanding and tasks undertaken. Registering or transferring details of CPD diary entries or certificates going back over the previous three-year period would help with the development of IT skills, and increase the confidence of those members who may be unsure of web-based systems.

The objectives section is the equivalent of the CPD planner, and although this would usually be completed prior to commencing CPD, in entering unplanned CPD at this initial stage, members are at least able to begin recording CPD online with a view to assessing objectives at a suitable point in due course.

Business skills illustration: RICS

In *Chartered Surveyor Monthly* (now *RICS Business*) of May 2002, the following commentary summarised the contribution that business skills can make to business growth. Margaret Harris of RICS was joint author.

In drawing primarily on technical skills, surveyors do not tend to possess a natural instinct in the fields of business and management.

RICS education and training initiatives, together with related corporate strategies, therefore seek to raise the profile and earning capacity of the profession through the development of business skills. What this means for individual members depends on their area of work, interest in business issues and overall career aspirations.

The ability to understand a client's business objectives not only ensures that a high-quality service is provided in individual cases, but can also lead to securing consultancy roles. In gaining new clients through a successful consultancy arm, instructions in mainstream casework may follow.

As well as advising such non-property companies on the strategic role that property can play, surveyors may help drive a property-related business forward, such as an investment, development or contracting company – or indeed a surveying practice. In fact, surveyors' detailed knowledge of their industry/market, combined with a basic appreciation of marketing and business development techniques, usually puts them a long way ahead of business and marketing consultants from non-property backgrounds.

Business skills encompass an understanding of economic and property market trends and, at a more advanced level, a knowledge of finance and accounting techniques. An up-to-date awareness of developments in information technology, including websites, e-mail and mobile communication, helps surveyors keep a step ahead of the competition, and for those with a keen business eye, also helps new opportunities to be exploited.

Effective management creates internal efficiencies in both the private and public sectors, and good interpersonal skills and presentation technique are essential for all professionals.

Business skills illustration: the profession's clients

From the RICS/Advantage West Midlands Annual Property Education Debate in 2003 in respect of the development of business skills within the surveying profession, *Estates Gazette*'s 'Mainly for Students' series reported as follows:

At the debate Mark Clarke of Advantage West Midlands illustrated how as a representative of one of the profession's major clients, he sees the Agency's more successful consultants drawing on business skills when delivering their commissions.

A key aspect highlighted was the ability of consultants to understand the strategic and operational aims of Advantage West Midlands, in relation to national and European planning and general policy issues when responding to commissions, and delivering clearly structured reports that provided a clear and consistent rationale to support the recommendations being made.

Project management skills were illustrated as another key area, involving the ability of a consultant to co-ordinate or participate within a multi-disciplinary team and deliver commissions to agreed timescales.

For mainly technical services, it was considered important that reports and presentations met essential business requirements, and were prepared with minimal input by the client.

Business skills illustration: surveying practices

This section is contributed by Scott Kind, GVA Grimley's Training and Development Manager. It formed part of the RICS/Advantage West Midlands Annual Property Education Debate in 2003, and is also now shown on RICS website.

The requirements of GVA Grimley's surveyors in respect of business skills, as with any company, vary – depending on their seniority within the firm, and the work in which they are involved.

Graduates, for example, will be immediately involved in client dealings, and must possess elementary interpersonal skills. It is also important that good relationships are developed with clients, and an understanding is gained of the operational business issues they face.

Understanding business issues

This could be a non-property business requiring property solutions which align with plans in respect of expansion or contraction, or a property investor looking to acquire new investment properties which fit well within their existing portfolio and the general aspirations of the business.

Surveyors would need to understand the contrast, for example, between the aspirations of predominantly family owned business to preserve family control, and being forced to grow relatively slowly, and a business attracting capital from numerous shareholders and other sources, and able to grow rapidly.

Easing business risk

It would not be in the surveyor's usual remit to provide accountancy and financial advice, but it would be important, for example, to highlight the risks that a client investor may face through property dealings, including the impact of poor investment performance on cash flow, asset values, gearing, balance sheet strength, ability to raise further finance etc.

For a non-property business, the effective use of property, occupationally and/or through leasing or sale (including sale and leaseback), could provide vital cash injection, but needs to be at acceptable risk and cost; such risks reflecting operational business issues as well as property factors.

Construction sector issues

Similarly, building surveyors will have to plan maintenance strategies on behalf of property owners/investors which have regard to affordability, cash flow, recoverability from tenants through leases, etc.

Project management work for construction projects bring softer business skills to the fore, again including interpersonal skills, but also leadership, team-building, negotiation and problem solving skills.

Technical skills

The above few examples (and there are many more) illustrate the difference between surveyors being primarily technically proficient, and surveyors being able to combine technical skills with business skills. The difference is reflected in the ability to undertake more challenging work – accounting for more complex business as well as property issues. It is also reflected in the contribution that surveyors can make to the fee earning capacity of their firm, and to their own salary and overall career potential.

It is important to stress that while GVA Grimley and other large practices seek graduates with business and management skills who will in due course help take the business forward, there are still key roles for surveyors who enjoy mainly the technical aspects of surveying.

People management, recruitment, internal financial issues

Irrespective of the extent to which surveyors are involved in technical areas or work more geared to business issues, surveyors will still usually require the skills to manage others, and get the best out of their team. Recruitment processes need to identify the best people, and an awareness is also required of wider areas such as employment law, business practices, etc.

It is also necessary for surveyors to understand aspects of their business such as fee structures, fee forecasting, and the profitability of individual fee lines.

Integrating training with RICS requirements

These elements are reflected in GVA Grimley's internal business and management skills training programmes. From September 2004, this will integrate, in part, with RICS post-graduate business qualification (although which is now under review at the time of publication).

GVA Grimley's policy for the qualification is already in place, and having restructured training accordingly, RICS requirements can be achieved cost-effectively.

Fast-track routes for young managers

At GVA Grimley, in order to identify the graduates of the calibre to take the business forward by securing managerial positions despite their young years, the policy in respect of the business qualification includes a fast-track option. Here, on immediately passing the APC, selected surveyors will begin a suitable qualification. For some, this will actually be a Diploma in Business Studies, or an MBA, both of which go beyond RICS 60 credit requirement.

Cross-departmental systems

GVA Grimley places importance on individual surveyors and departments being aware of the work of surveyors in other disciplines/departments. In addition to this helping clients receive good quality, comprehensive, advice, effective cross-departmental systems also help wider fee earning opportunities to be identified for wider benefit of the GVA Grimley business.

Anticipating market developments

New fee earning opportunities are also identified by surveyors through their understanding of property market and wider business trends, and their anticipation of the effects of developments in technology, law, market practice, etc.

The rapid growth and profitability of GVA Grimley's telecoms team is a good example, as in fact, is the ease at which the telecoms team were able to contract when growth in the sector stalled.

Senior surveyors taking business qualifications

As an example of senior surveyors also taking business qualifications, Colin Sharp, a senior partner and a main board member of GVA Grimley, recently completed a masters degree in business administration at City University. Within previous 'RICS Relaunch – Raising the standards' coverage, Colin commented that, 'the rapid expansion of our telecoms team to capitalise on the sector's growth highlighted two vital components: our own effective business planning, and an understanding of clients' operational objectives'.

Pitches for new work

Another area in which GVA Grimley's more senior surveyors are involved is pitches for new business. Here, in searching for an edge in terms of service quality and/or fee levels, it is important to be able to identify prospective clients' key requirements. As with mainstream areas of surveying, strong verbal and written presentation skills are required, together with an ability to demonstrate a good track record, and the high calibre of the surveyors likely to be involved in the work.

Structuring the UK business

GVA Grimley's senior partners are responsible for the structuring of the UK business. This involves 16 national business units covering the range of surveying disciplines which make up GVA Grimley's range of consultancy services spread across 11 offices. The structure combines centralised functions of leadership and business strategy with autonomy within business units at local level, thus enabling individual teams to develop their client base and range of services (and also receive the due personal rewards). Flexibility within the structures enables account to be taken of factors such as variations in client work throughout the offices, specialist teams being located in certain offices, and staff numbers varying between offices.

UK and global growth plans

Senior partners of GVA Grimley and GVA Worldwide also have to understand global business trends, international property market factors and the operational qualities of property businesses when assessing the potential for acquisitions and mergers. Here, chartered surveyors combine an understanding of property and business to ensure that any new businesses provide the right strategic fit with GVA Grimley's existing operations. This is in terms of many factors, including the respective specialisms and typical client base of each firm, the extent to which services may overlap, the extent to which combined forces help open up new markets, the ease of integrating staff, and whether cultures and working practices are compatible.

Working cultures

For all members of staff, the right work-life balance is important in ensuring optimum productivity, which for both the firm and individual, mean that high quality work undertaken in the most expedient time-frame, yields the greater due financial rewards.

Team-based approaches

Although this illustration focuses on the work of surveyors, is important to mention the requirement for business and managerial skills among property administrative staff, and other support roles, including human resources. Here, effective, team-based, support to fee earners, is vital in ensuring that the business operates efficiently.

Further information

Further information is available at:

www.rics.org/careers/cpd/post_qual.html
www.rics.org/careers/cpd/business_skills.html
www.rics.org/careers/cpd/success_stories.html

These are correct at the time of publication.

Index